MATERIALS SCIENCE

This text is intended for a second-level course in materials science and engineering. Chapters encompass crystal symmetry including quasi-crystals and fractals, phase diagrams, diffusion including treatment of diffusion in two-phase systems, solidification, solid-state phase transformations, amorphous materials, and bonding in greater detail than is usual in introductory materials science courses. Additional subject material includes stereographic projection, the Miller–Bravais index system for hexagonal crystals, microstructural analysis, the free energy basis for phase diagrams, surfaces, sintering, order–disorder reaction, liquid crystals, molecular morphology, magnetic materials, porous materials, and shape memory and superelastic materials. The final chapter includes useful hints in making engineering calculations. Each chapter has problems, references, and notes of interest.

William F. Hosford is a Professor Emeritus of Materials Science and Engineering at the University of Michigan. Professor Hosford is the author of a number of books including the leading selling *Metal Forming: Mechanics and Metallurgy*, 2/e (with R. M. Caddell), *Mechanics of Crystals and Textured Polycrystals*, *Physical Metallurgy*, and *Mechanical Behavior of Materials*.

Materials Science

AN INTERMEDIATE TEXT

WILLIAM F. HOSFORD
University of Michigan

CAMBRIDGE
UNIVERSITY PRESS

32 Avenue of the Americas, New York NY 10013-2473, USA

Cambridge University Press is part of the University of Cambridge.

It furthers the University's mission by disseminating knowledge in the pursuit of education, learning and research at the highest international levels of excellence.

www.cambridge.org
Information on this title: www.cambridge.org/9780521867054

© William F. Hosford 2007

This publication is in copyright. Subject to statutory exception
and to the provisions of relevant collective licensing agreements,
no reproduction of any part may take place without the written
permission of Cambridge University Press.

First published 2007

A catalogue record for this publication is available from the British Library

Library of Congress Cataloguing in Publication data

Hosford, William F.
Materials science : an intermediate text / William F. Hosford.
   p. cm.
Includes bibliographical references and index.
ISBN-13: 978-0-521-86705-4 (hardback)
ISBN-10: 0-521-86705-3 (hardback)
1. Materials science – Textbooks.   I. Title.
TA403.H63   2006
620.1'1 – dc22   2006011097

ISBN 978-0-521-86705-4 Hardback

Cambridge University Press has no responsibility for the persistence or accuracy of URLs for external or third-party internet websites referred to in this publication, and does not guarantee that any content on such websites is, or will remain, accurate or appropriate.

Contents

Preface	*page* xiii

1 Microstructural Analysis ... 1

Grain size	1
Relation of grain boundary area per volume to grain size	3
Relation of intersections per area and line length	4
Volume fraction of phases	4
Alloy composition from volume fraction of two or more phases	4
Microstructural relationships	5
Three-dimensional relations	6
Kelvin tetrakaidecahedron	6
Notes of interest	8
References	8
Problems	9

2 Symmetry ... 11

Crystal systems	11
Space lattices	11
Quasicrystals	14
Fractals	17
Note of interest	18
References	19
Problems	19

3 Miller–Bravais Indices for Hexagonal Crystals ... 21

Planar indices	21
Direction indices	22
Three-digit system	23
Note of interest	24
References	24
Problems	24

4 Stereographic Projection 26

Projection 26
Standard cubic projection 27
Locating the $hk\ell$ pole in the standard stereographic projection of a cubic crystal 28
Standard hexagonal projection 30
Spherical trigonometry 31
Note of interest 31
References 31
Problems 31

5 Crystal Defects 33

Vacancies in pure metals 33
Point defects in ionic crystals 34
Dislocations 36
Burgers vectors 37
Energy of dislocations 38
Stress fields around dislocations 38
Partial dislocations 39
Notes of interest 40
References 41
Problems 41

6 Phase Diagrams 43

The Gibbs phase rule 43
Invariant reactions 44
Ternary phase diagrams 44
Notes of interest 49
References 49
Problems 50

7 Free Energy Basis for Phase Diagrams 52

Gibbs free energy 52
Enthalpy of mixing 52
Entropy of mixing 53
Solid solubility 55
Relation of phase diagrams to free energy curves 55
Pressure effects 57
Metastability 57
Extrapolations of solubility limits 60
Notes of interest 61
References 62
Problems 62

8 Ordering of Solid Solutions ... 64
Long-range order — 64
Effect of long-range order on properties — 67
Short-range order — 67
Note of interest — 67
References — 68
Problems — 68

9 Diffusion ... 69
Fick's first law — 69
Fick's second law — 70
Solutions of Fick's second law and the error function — 70
Mechanisms of diffusion — 73
Kirkendall effect — 74
Temperature dependence — 75
Special diffusion paths — 76
Darken's equation — 77
Diffusion in systems with more than one phase — 78
Note of interest — 81
References — 82
Problems — 82

10 Freezing ... 85
Liquids — 85
Homogeneous nucleation — 85
Heterogeneous nucleation — 88
Growth — 89
Grain structure of castings — 90
Segregation during freezing — 91
Zone refining — 93
Steady state — 95
Dendritic growth — 95
Gas solubility and gas porosity — 98
Growth of single crystals — 98
Eutectic solidification — 98
Peritectic freezing — 100
Notes of interest — 101
References — 101
Problems — 102

11 Phase Transformations ... 104
Nucleation in the solid state — 104
Eutectoid transformations — 106

Avrami kinetics 108
Growth of precipitates 111
Transition precipitates 113
Precipitation-free zones 113
Ostwald ripening 113
Martensitic transformations 114
Spinodal decomposition 116
Note of interest 118
References 119
Problems 119

12 Surfaces 121
Relation of surface energy to bonding 121
Orientation-dependence of surface energy 122
Surfaces of amorphous materials 125
Grain boundaries 125
Segregation to surfaces 127
Direct measurements of surface energy 128
Measurements of relative surface energies 129
Wetting of grain boundaries 130
Relative magnitudes of energies 131
Note of interest 131
References 131
Problems 131

13 Bonding 133
Ionic binding energy 133
Melting points 134
Elastic moduli 134
Covalent bonding 136
Geometric considerations 136
Ionic radii 139
Structures of compounds 140
Note of interest 142
References 143
Problems 143

14 Sintering 144
Mechanisms 144
Early stage of sintering 146
Intermediate stage of sintering 147
Final stage of sintering 147
Loss of surface area 147
Particle-size effect 148

	Activated sintering	150
	Liquid-phase sintering	150
	Hot isostatic pressing	151
	Note of interest	151
	References	151
	Problems	151
15	**Amorphous Materials**	**153**
	Glass transition	153
	Glass transition in polymers	154
	Molecular length	154
	Hard sphere model	155
	Voronoi cells	157
	Silicate glasses	157
	Chemical composition	158
	Bridging versus nonbridging oxygen ions	158
	Glass viscosity	159
	Thermal shock	160
	Thermal expansion	161
	Vycor	161
	Devitrification	162
	Delayed fracture	163
	Other inorganic glasses	163
	Metal glasses	164
	Note of interest	166
	References	167
	Problems	167
16	**Liquid Crystals**	**168**
	Types of liquid crystals	168
	Orientational order parameter	169
	Disclinations	170
	Lyotropic liquid crystals	171
	Temperature and concentration effects	171
	Phase changes	172
	Optical response	173
	Liquid crystal displays	174
	Note of interest	174
	References	175
	Problems	175
17	**Molecular Morphology**	**176**
	Silicates	176
	Molybdenum disulfide	178

	Carbon: graphite	179
	Diamond	179
	Carbon fibers	180
	Fullerenes	180
	Nanotubes	181
	Zeolites	182
	Notes of interest	183
	References	183
	Problems	183
18	**Magnetic Behavior of Materials**	**184**
	Ferromagnetism	184
	Exchange energy	185
	Magnetostatic energy	187
	Magnetocrystalline energy	188
	Magnetostrictive energy	189
	Physical units	189
	The B–H curve	190
	Curie temperature	191
	Bloch walls	191
	Magnetic oxides	192
	Soft versus hard magnetic materials	194
	Soft magnetic materials	194
	Hard magnetic materials	197
	Square-loop materials	199
	Notes of interest	200
	References	201
	Problems	201
19	**Porous and Novel Materials**	**202**
	Applications of porous materials	202
	Fabrication of porous foams	202
	Morphology of foams	203
	Relative density of foams	203
	Structural mechanical properties	204
	Honeycombs	204
	Novel structures	205
	Notes of interest	205
	Reference	206
	Problems	206
20	**Shape Memory and Superelasticity**	**208**
	Shape memory alloys	208
	Superelasticity	209

	Applications	212
	Shape memory in polymers	212
	Note of interest	213
	References	213
	Problems	213
21	**Calculations**	**214**
	Estimates	214
	Sketches	215
	Units	217
	Available data	219
	Algebra before numbers	220
	Ratios	220
	Percentage changes	221
	Finding slopes of graphs	221
	Log-log and semilog plots	222
	Graphical differentiation and integration	224
	Iterative and graphical solutions	226
	Interpolation and extrapolation	228
	Analyzing extreme cases (bounding)	228
	Significant figures	229
	Logarithms and exponents	230
	The Greek alphabet	231
	Problems	231
	Index	235

Preface

This text is written for a second-level materials science course. It assumes that the students have had a previous course covering crystal structures, phase diagrams, diffusion, Miller indices, polymers, ceramics, metals, and other basic topics. Many of those topics are discussed in further depth, and new topics and concepts are introduced. The coverage and order of chapters are admittedly somewhat arbitrary. However, each chapter is more or less self-contained so those using this text may omit certain topics or change the order of presentation.

The chapters on microstructural analysis, crystal symmetry, Miller–Bravais indices for hexagonal crystals, and stereographic projection cover material that is not usually covered in introductory materials science courses. The treatment of crystal defects and phase diagrams is in greater depth than the treatments in introductory texts. The relation of phase diagrams to free energy will be entirely new to most students. Although diffusion is covered in most introductory texts, the coverage here is deeper. It includes the Kirkendall effect, Darken's equation, and diffusion in the presence of two phases.

The topics of surfaces and sintering will be new to most students. The short chapter on bonding and the chapters on amorphous materials and liquid crystals introduce new concepts. These are followed by treatment of molecular morphology. The final chapters are on magnetic materials, porous and novel materials, and the shape memory.

This text may also be useful to graduate students in materials science and engineering who have not had a course covering these materials.

The author wishes to thank David Martin for help with liquid crystals.

1 Microstructural Analysis

Many properties of materials depend on the grain size and the shape of grains. Analysis of microstructures involves interpreting two-dimensional cuts through three-dimensional bodies. Of interest are the size and aspect ratios of grains, and the relations between grain size and the amount of grain boundary area per volume. Also of interest is the relation between the number of faces, edges, and corners of grains.

Grain size

There are two commonly used ways of characterizing the grain size of a crystalline solid. One is the ASTM grain size number, N, defined by

$$n = 2^{N-1} \quad \text{or} \quad N = 1 + \ln(n)/\ln 2, \tag{1.1}$$

where n is the number of grains per square inch observed at a magnification of 100X. Large values of N indicate a fine grain size. With an increase of the grain diameter by a factor of $\sqrt{2}$, the value of n is cut in half and N is decreased by 1.

EXAMPLE 1.1. Figure 1.1 is a micrograph taken at 200X. What is the ASTM grain size number?

SOLUTION: There are 29 grains entirely within the micrograph. Counting each grain on an edge as one half, there are $22/2 = 11$ edge grains. Counting each corner grain as one quarter, there is 1 corner grain. The total number of grains is 41. The 12 square inches at 200X would be 3 square inches at 100X, so $n = 41/3 = 13.7$. From Equation (1.1),

$$N = \ln(n)/\ln(2) + 1 = 4.78 \text{ or } 5.$$

The average linear intercept diameter is the other common way to characterize grain sizes. The system is to lay down random lines on the microstructure and count the number of intersections per length of line. The average intercept diameter is then $\bar{\ell} = L/N$, where L is the total length of line and N is the number

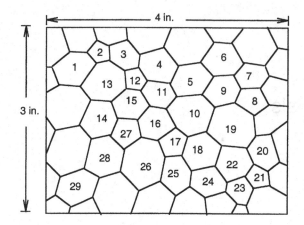

1.1. Counting grains in a microstructure at 200X.

of intercepts. Alternatively, a rectangular grid of lines may be laid down on an equiaxed microstructure.

EXAMPLE 1.2. Find the average intercept diameter for the micrograph in Figure 1.1.

SOLUTION: In Figure 1.2, $6 \times 4 + 5 \times 3 = 39$ inches of line are superimposed on the microstructure. This corresponds to $(39 \text{ in.}/200)(25.4 \text{ mm/in.}) = 4.95$ mm. There are 91 intercepts so $\bar{\ell} = .495/91 = 0.054$ mm $= 54$ µm.

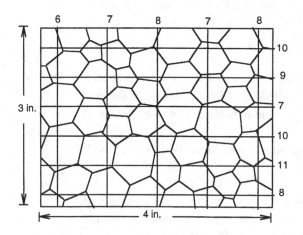

1.2. Finding the linear intercept grain size of a microstructure at 200X.

For random microstructures, $\bar{\ell}$ and the ASTM grain size are related. An approximate relationship can be found by assuming that the grains can be approximated by circles of radius, r. The area of a circular grain, πr^2, can be expressed as the average linear intercept, $\bar{\ell}$, times its width, $2r$, as shown in Figure 1.3, so $\bar{\ell} \cdot 2r = \pi r^2$. Therefore,

$$r = (2/\pi)\bar{\ell} \quad \text{or} \quad \bar{\ell} = (\pi/2)r. \tag{1.2}$$

MICROSTRUCTURAL ANALYSIS

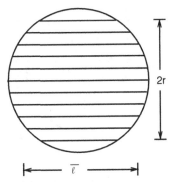

1.3. The area of a circle, πr^2, equals the average intercept times twice the radius, $2\bar{\ell}r$, so $\bar{\ell} = (\pi/2)r$.

Thus, the area per grain is $A = 2r\bar{\ell} = (4/\pi)\bar{\ell}^2$. The number of grains per area is $(\pi/4)/\bar{\ell}^2$. From the definition of n, the number of grains per area is also $n[(25.4\,\text{mm/in.})/(100\,\text{in.})]^2$. Substituting $n = 2^{N-1} = 2^N/2$ and equating,

$$(\pi/4)/\bar{\ell}^2 = (2^N/2)(0.254)^2. \tag{1.3}$$

Solving for $\bar{\ell}$,

$$\bar{\ell} = 4.93/2^{N/2}. \tag{1.4}$$

Often grains are not equiaxed. They may be elongated in the direction of prior working. Restriction of grain growth by second-phase particles may prevent formation of equiaxed grains by recrystallization. In these cases, the linear intercept grain size should be determined from randomly oriented lines or an average of two perpendicular sets of lines. The degree of shape anisotropy can be characterized by an aspect ratio, α, defined as the ratio of average intercept in the direction of elongation to that at 90°:

$$\alpha = \bar{\ell}_\parallel/\bar{\ell}_\perp. \tag{1.5}$$

Relation of grain boundary area per volume to grain size

The grain boundary area per volume is related to the linear intercept. Assuming that grain shapes can be approximated by spheres, the grain boundary surface per grain is $2\pi R^2$, where R is the radius of the sphere. (The reason that it is not $4\pi R^2$ is that each grain boundary is shared by two neighboring grains.) The volume per spherical grain is $(4/3)\pi R^3$, so the grain boundary area/volume, S_v, is given by

$$S_v = (2\pi R^2)/[(4/3)\pi R^3] = 3/(2R). \tag{1.6}$$

To relate the spherical radius, R, to the linear intercept, $\bar{\ell}$, consider the circle through its center, which has an area of πR^2 (Figure 1.4). The volume equals the product of this area, πR^2, and the average length of line, $\bar{\ell}$, perpendicular to it, $v = \bar{\ell}\pi R^2$. Therefore, $(4/3)\pi R^3 = \pi R^2 \bar{\ell}$ or $R = (3/4)\bar{\ell}$. Substituting into $S_v = 3/(2R)$,

$$S_v = 2/\bar{\ell}. \tag{1.7}$$

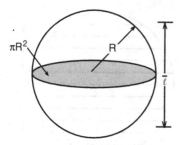

1.4. The volume of a sphere $= \bar{\ell}\pi R^2$.

Relation of intersections per area and line length

The number of intersections per area of dislocations with a surface is less than the total length of dislocation line per volume. Consider a single line of length L in a box of height h and area of A. The number of intersections per area, N_A, equals $1/A$ (Figure 1.5). The length per volume is $L_V = L/(hA)$ so $N_A/L_V = h/L$. Because $\cos\theta = h/L$, $N_A/L_V = \cos\theta$. For randomly oriented lines, the number oriented between θ and $\theta + d\theta$ is $dn = ndf$, where $df = \sin\theta d\theta$. For randomly oriented lines, $N_A/L_V = \int^{2\pi} \cos\theta \sin\theta d\theta = 1/2$. Therefore,

$$N_A = L_V/2. \qquad (1.8)$$

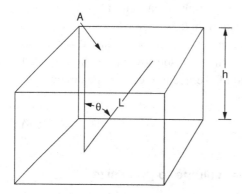

1.5. Relation of the number of intersections per area with the length of line per volume.

Volume fraction of phases

Point counting is the easiest way of determining the volume fraction of two or more phases in a microstructure. The volume fraction of a phase equals the fraction of points in an array that lies on that phase. A line count is another way of finding the volume fraction. If a series of lines are laid on a microstructure, the volume fraction of a phase equals the fraction of the total line length that lies on that phase.

Alloy composition from volume fraction of two or more phases

The composition of an alloy can be found from the volume fractions of phases. The relative weight of component B in the α phase is $(V_\alpha)(\rho_\alpha)(C_\alpha)$, where V_α is

MICROSTRUCTURAL ANALYSIS

the volume fraction of α, ρ_α is the density of α, and C_α is the composition (%B) of the α phase. With similar expressions for the other phases, the relative weight of component B, W_B, is given by

$$W_B = (V_\alpha)(\rho_\alpha)(C_\alpha) + (V_\beta)(\rho_\beta)(C_\beta) + \cdots \quad (1.9)$$

With similar expressions for the other components, the overall composition of the alloy is

$$\%B = 100 W_B/(W_A + W_B + \cdots). \quad (1.10)$$

Microstructural relationships

Microstructures consist of three-dimensional networks of cells or grains that fill space. Each cell is a polyhedron with faces, edges, and corners. Their shapes are strongly influenced by surface tension. However, before examining the nature of three-dimensional microstructures, the characteristics of two-dimensional networks will be treated.

A two-dimensional network of cells consists of polygons, edges (sides), and corners. The number of each is governed by the simple relation

$$P - E + C = 1, \quad (1.11)$$

where P is the number of polygons, E is the number of edges, and C is the number of corners. Figure 1.6 illustrates this relationship. If the microstructure is such that three and only three edges meet at each corner, $E = (3/2)C$, so

$$P - C/2 = 1 \quad \text{and} \quad P - E/3 = 1. \quad (1.12)$$

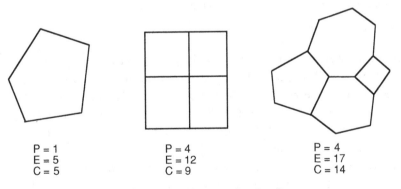

1.6. Three networks of cells illustrating that $P - E + C = 1$.

For large numbers of cells, the one on the right-hand side of Equations (1.9) and (1.10) becomes negligible, so $E = 3P$ and $C = 2P$. This restriction of three edges meeting at a corner also requires that the average angle at which the edges meet is 120° and that the average number of sides per polygon is six.

If the edges were characterized by a line tension (in analogy to the surface tension of surfaces in three dimensions) and if the line tensions for all edges were equal, equilibrium would require that all edges meet at 120°, so cells with more than six edges would have to be curved with the center of curvature away from the cell and those cells with fewer than six sides would be curved the opposite way, as shown in Figure 1.7. Since boundaries tend to move toward their centers of curvature, the cells with large numbers of sides would tend to grow and those with few sides should shrink. Only a network in which all of the cells were regular hexagons would be stable.

1.7. The sides of grains with fewer than six neighbors are inwardly concave (left). The sides of grains with more than six neighbors are outwardly concave (right).

Three-dimensional relations

Euler proposed that for a single body

$$C - E + F - B = 1. \tag{1.13}$$

For an infinite array of three-dimensional bodies,

$$C - E + F - B = 0. \tag{1.14}$$

Here, B is the number of bodies (grains), F is the number of faces, E is the number of edges, and C is the number of corners. Consider an isolated cube, for example. There is one body, and there are six faces, 12 edges, and eight corners. $B = 1$, $F = 6$, $E = 12$, and $C = 8$. $8 - 12 + 6 - 1 = 1$. For an infinite array of stacked cubes, each face is shared by two cubes so $F = 6B/2$. Each edge is shared by four cubes so $E = 8B/4$, and each corner is shared by eight cubes so $C = 12B/8$. Substituting into Euler's equation, $8B/8 - 12B/4 + 6B/2 - B = 0$. Table 1.1 illustrates Equation (1.11) for several simple polyhedra.

Kelvin tetrakaidecahedron

Grains in a real material have certain restrictions: Each corner is shared by four grains, and each edge is shared by three grains. Furthermore, grains stack in such a way as to fill space. Very few simple shapes fulfill these conditions. One simple

MICROSTRUCTURAL ANALYSIS

Table 1.1. Characteristics of several polyhedra

Polyhedron	Faces (F)	Edges (E)	Corners (C)	F − E + C
Tetrahedron	4	6	4	2
Cube	6	12	8	2
Octahedron	8	12	6	2
Dodecahedron(cubic)	12	24	14	2
Dodecahedron(pentag.)	12	30	20	2
Tetrakaidecahedron	14	36	24	2

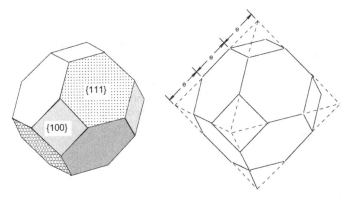

1.8. The Kelvin tetrakaidecahedron and its construction by truncation of an octahedron by a cube. The edges of the tetrakaidecahedron are one third as long as the edges of the octahedron.

shape is the tetrakaidecahedron proposed by Lord Kelvin.* Figure 1.8 shows that it can be thought of as a cube with each corner truncated by an octahedron. Alternatively, it can be thought of as an octahedron with each corner truncated by a cube. There are 14 faces, 36 edges, and 24 corners. For an infinite array of these polyhedra,

$$F = 14B/2 = 7B, C = 24B/4 = 6B, \quad \text{and} \quad E = 36B/3 = 12B,$$
$$\text{so } C - E + F - B = 6B - 12B + 7B - B = 0.$$

This shape is a useful approximation for analyzing grains in a polycrystal. For example, calculation of the surface area of the faces to the grain volume can be compared with other solid shapes and a sphere. Six of these are squares parallel to $\{100\}$ planes and eight are regular hexagons parallel to $\{111\}$ planes. There are 24 corners and 36 edges. Thus, the total length of edges is $36e$, where e is the length of an edge, and the total surface area is the area of the six square faces plus the eight hexagonal faces:

$$6e^2 + 8(3\sqrt{3})e^2 = 47.569e^2.$$

The volume is the volume of the octahedron less the volume of the six truncated pyramids:

$$[9\sqrt{2} - 6(1/3\sqrt{2})]e^3 = 8\sqrt{2}e^3 = 11.314e^3.$$

---

* W. T. Lord Kelvin, *Proc. R. Soc.* 55 (1894).

Of the 14 faces, 6 have four edges and 8 have six edges. The average number of edges per face is $(6 \times 4 + 8 \times 6)/14 = 5\frac{1}{7}$. This is very close to the results of experiments on β brass, vegetable cells, and soap bubbles, as shown in Figure 1.9. For the Kelvin tetrakaidecahedron the ratio of surface area to that of a sphere of the same volume is 1.099. Most other shapes have much higher ratios.

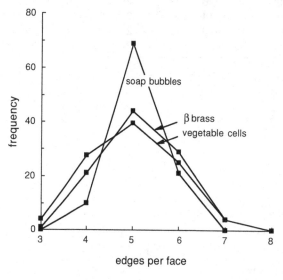

1.9. Frequency of polygonal faces with different numbers of edges. Data from C. S. Smith, in *Metal Interfaces* (Cleveland, OH: ASM, 1952). Reprinted with permission from ASM International®. All rights reserved. www.asminternational.org.

NOTES OF INTEREST

1. Lord Kelvin (1824–1907), a Scottish mathematician and physicist, did the pioneering work on the second law of thermodynamics, arguing that it was the explanation of irreversible processes. He noted that the continual increase of entropy would lead to a universe with a uniform temperature and maximum entropy.
2. Waire and Phelan* report that space filling is 0.3% more efficient with an array of of six polyhedra with 14 faces and two polyhedra with 12 faces than with the Kelvin tetrakaidecahedron. (This calculation allows faces in each to be curved.) The 14-faced polyhedra have 12 pentagonal and 2 hexagonal faces, while the 12-faced polyhedra have distorted pentagons for faces. The average number of faces per polyhedra $= (6 \times 14 + 2 \times 12)/8 = 13.5$.

REFERENCES

R. T. DeHoff and F. N. Rhine, eds. *Quantitative Metallography*. New York: McGraw-Hill, 1968.
W. T. Lord Kelvin. *Phil. Mag.* 24 (1887): 503–14.
C. S. Smith. In *Metal Interfaces*, pp. 65–113. Cleveland, OH: ASM, 1952.
E. E. Underwood. *Quantitative Stereology*. Boston: Addison-Wesley, 1970.

* D. L. Weaire and R. Phelan. *Phil. Mag., Letters* 87 (1994): 345–50.

MICROSTRUCTURAL ANALYSIS

PROBLEMS

1. A soccer ball has 32 faces. They are all either pentagons or hexagons. How many are pentagons?

2. Figure 1.10 is a microstructure at a magnification of 200X.
   A. Determine the ASTM grain size number.
   B. Determine the intercept grain size.
   C. Compare the answers to A and B using Equation (1.5).

3. Count the number of triple points in Figure 1.10 and deduce the ASTM grain size from this count. Compare with your answer to Problem 2A.

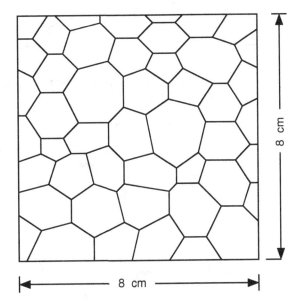

1.10. Hypothetical microstructure at a magnification of 200X.

4. What is the linear intercept grain size (in millimeters) corresponding to an ASTM grain size number of 8?

5. Dislocation density is often determined by counting the number of dislocations per area intersecting a polished surface. If the dislocation density in cold-worked copper is found to be $2 \times 10^{10}/cm^2$, what is the total length of dislocation line per volume?

6. Calculate the average number of edges per face for the space-filling array of polyhedra reported by Waire and Phelan.

7. If the ASTM grain size number is increased by one, by what factor is the number of grains per volume changed?

8. If a material with grains shaped like tetrakaidecahedra were recrystallized and new grains were nucleated at each corner, by what factor would the grain diameter, $\bar{\ell}$, change?

9. Derive an equation relating the aspect ratio of a microstructure after uniaxial tension to the strain, assuming that the microstructure was initially equiaxed.

10. Determine the aspect ratio in Figure 1.11.

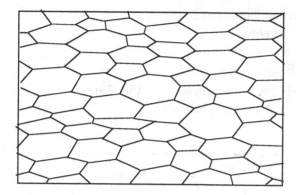

1.11. Microstructure for Problem 10.

11. Determine the volume fraction of graphite in the cast iron shown in Figure 1.12.

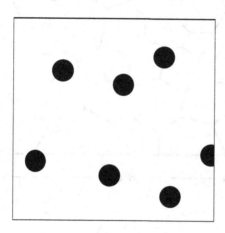

1.12. A schematic drawing of ferritic ductile cast iron. The white areas are ferrite and the dark circles are graphite.

2 Symmetry

Crystal systems

Crystals can be classified into seven systems. A crystal system is defined by the repeat distances along its axes and the angles between its axes. Table 2.1 lists the seven systems.

Table 2.1. The seven crystal systems

System	Axial lengths	Axial angles
Triclinic	$a \neq b \neq c$	$\alpha \neq \beta \neq \gamma \neq 90°$
Monoclinic	$a \neq b \neq c$	$\alpha = \beta = 90° \neq \gamma$
Orthorhombic	$a \neq b \neq c$	$\alpha = \beta = \gamma = 90°$
Tetragonal	$a = b \neq c$	$\alpha = \beta = \gamma = 90°$
Cubic	$a = b = c$	$\alpha = \beta = \gamma = 90°$
Rhombohedral	$a = b = c$	$\alpha = \beta = \gamma \neq 90°$
Hexagonal	$a = b \neq c$	$\alpha = \beta = 90°, \gamma = 120°$

These systems can be described in terms of their symmetry elements. A triclinic crystal has only a center of symmetry. Monoclinic crystals have a single axis of twofold rotational symmetry. Orthorhombic crystals have three mutually perpendicular axes of twofold symmetry. With tetragonal symmetry, there is a single axis of fourfold symmetry. Cubic crystals are characterized by four threefold axes of symmetry, the <111> axes. There is a single axis of threefold symmetry in the rhombohedral system. The hexagonal system involves a single axis of sixfold symmetry.

Space lattices

Crystals can be further divided into 14 space lattices, which describe the positions of lattice points. For example, there are three cubic space lattices. The simple cubic has lattice points only at the corners of the cubic cell; the body-centered cubic (bcc) has lattice points at the corners and the body-centering position, and the face-centered cubic (fcc) has lattice points at the corners and the centers of the faces. Table 2.2 and Figure 2.1 illustrate these.

12 MATERIALS SCIENCE: AN INTERMEDIATE TEXT

Table 2.2. The 14 space lattices

Triclinic	Simple tetragonal
Simple monoclinic	Body-centered tetragonal
Base-centered monoclinic	Simple cubic
Simple orthorhombic	Body-centered cubic
Body-centered orthorhombic	Face-centered cubic
Base-centered orthorhombic	Rhombohedral
Face-centered orthorhombic	Hexagonal

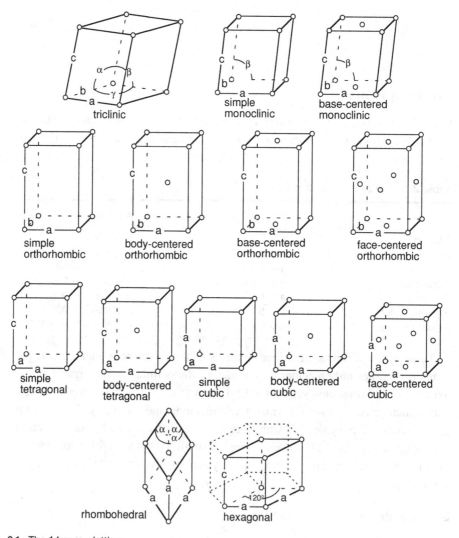

2.1. The 14 space lattices.

Symmetry elements include axes of twofold, threefold, fourfold, and sixfold rotational symmetry and mirror planes. There are also axes of rotational inversion symmetry. With these, there are rotations that cause mirror images. For example, a simple cube has three <100> axes of fourfold symmetry, four axes of <111>

SYMMETRY

threefold symmetry, and six <110> axes of twofold symmetry. A cube also has nine mirror planes (three {100} planes and six {110} planes). See Figure 2.2.

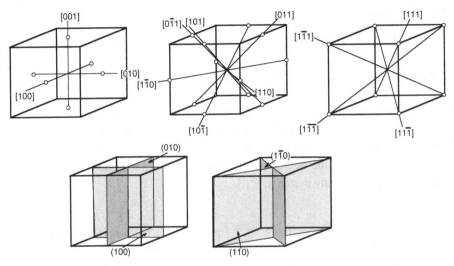

2.2. Symmetry elements of a cube. There are three fourfold axes, four threefold axes, and six twofold axes of rotation. There are three {100} mirror planes and six {110} mirror planes (only two are shown).

On the other hand, not all crystals with a cubic space lattice have all of the symmetry elements. Consider a tetrahedron (Figure 2.3). It has four axes of threefold symmetry, but the <100> directions have only twofold symmetry. There is no mirror symmetry about the {100} planes, but the six {110} planes do have mirror symmetry.

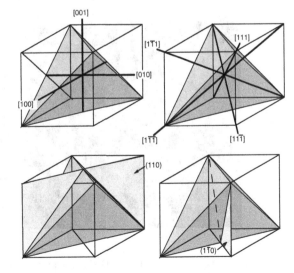

2.3. Symmetry elements of a tetrahedron. The {101}, {10$\bar{1}$}, {011}, and {01$\bar{1}$} planes also have mirror symmetry.

There are 32 crystal classes that describe all of the possible combinations of crystal systems and symmetry elements. These are treated in other texts.

Quasicrystals

There are a number of polyhedra that have axes of fivefold symmetry (Figure 2.4). However, there are no crystal classes or space lattices that permit fivefold symmetry. In 1984, Schectman et al.* found that the diffraction patterns from an aluminum–manganese alloy showed apparent tenfold symmetry (Figure 2.5). Figure 2.6 is a scanning electron microscope (SEM) photograph of a grain of $Al_{62}Cu_{25.5}Fe_{12.5}$ that shows fivefold symmetry. *Quasicrystals* are composed of certain combinations of polyhedra that fill space and have apparent five- or tenfold symmetry with some degree of short-range order. Such quasicrystals have since been found in many systems.

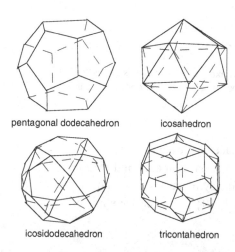

2.4. *Opposite:* Several polyhedra with axes of fivefold symmetry.

2.5. *Below, left:* Diffraction pattern from an aluminum–manganese alloy showing apparent tenfold symmetry. From C. Janot, *Quasicrystals, A Primer*, 2nd ed. (London: Oxford Univ. Press, 1994), p. 102, top photo (a).

2.6. *Below, right:* Scanning electron microscope (SEM) photograph of a dodecahedral grain of $Al_{62}Cu_{25.5}Fe_{12.5}$ showing fivefold symmetry. From C. Janot, *Quasicrystals, A Primer*, 2nd ed. (London: Oxford Univ. Press, 1994), p. 86, bottom photo (c).

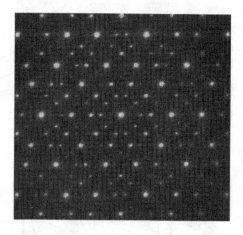

* D. Schechtman, I. Bloch, D. Gratias, and J. W. Cahn, *Phys. Rev. Lett.* 53 (1984): 1951–3.

SYMMETRY

A crystal has both symmetry and long-range order. It also has translational order; it can be replicated by small translations. It is possible to have both symmetry and long-range order without translational order. A one-dimensional example is a Fibonacci series that is composed of two segments, A and B. The series consists of terms N_n such that $N_n = N_{n-1} + N_{n-2}$. For example, the series starting with A and B is BA, BAB, BABBA, BABBABAB, ... Such a series has long-range order and will not repeat itself if N_{n-2}/N_{n-1} is an irrational number. For the series starting 0, 1, $N_{n-2}/N_{n-1} \to \tau = (1 + \sqrt{5})2$, which is called the *golden ratio*.

Penrose tiling patterns are two-dimensional analogs of quasicrystals. They fill space with patterns that have no long-range order. They require tiles of at least two different shapes. Figure 2.7 illustrates two shapes that can be tiled to produce patterns with fivefold short-range order. The interior angles are multiple integers of $\pi/10$. Figure 2.8 shows such a two-dimensional pattern. Note that there is fivefold rotational symmetry about the dark point in the center. However, there is no other point about which there is fivefold symmetry, even if the tiling is extended indefinitely.

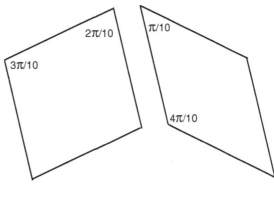

2.7. Two two-dimensional tiles that can be assembled into a tiling pattern with short-range fivefold symmetry.

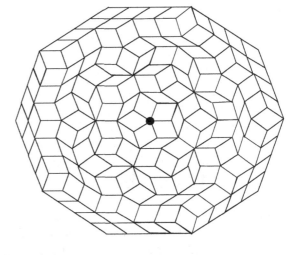

2.8. Penrose tiling with a tile having a 36° interior angle and another with a 72° interior angle.

If the tiling in Figure 2.8 is rotated 36° about the fivefold axis and translated the right amount, there is coincidence of the vertices with those of the original tiling though there is no longer a center of symmetry. This 36° rotation corresponds to tenfold symmetry.

A Fibonacci series in which each element of the series is the sum of the previous two elements is a one-dimensional analog. An example is the series, starting with L and S,

L S SL SLS SLSSL SLSSLSLS....

If $L/S = \tau = 2\cos 36° = (1 + \sqrt{5})/2 = 1.618034$, which is the golden ratio, the sequence has no repetition but there are diffraction peaks.

Certain polyhedra (Figure 2.9) can be assembled into a three-dimensional tiling to form a quasicrystal with regimes of icosahedral symmetry. An icosahedron has 20 faces and 12 axes of fivefold symmetry, as shown in Figure 2.10. The structures of $MoAl_{12}$ and WAl_{12} can be described as clusters of 12 aluminum atoms around molybdenum (or tungsten) atoms forming icosahedrons that fill space in a bcc arrangement.

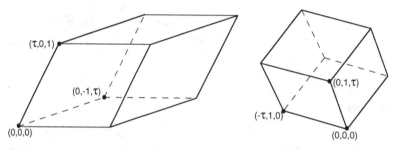

2.9. Oblate and prolate rhombohedrons that can be combined to form three-dimensional tiling necessary for a quasicrystal.

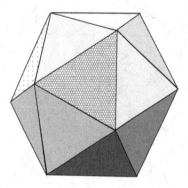

2.10. An icosahedron with 20 faces and six axes of fivefold symmetry.

An electron diffraction pattern of the aluminum–manganese alloy and a computed Fourier pattern of a three-dimensional Penrose tiling are shown in Figure 2.11.

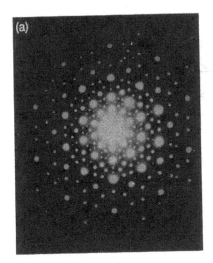 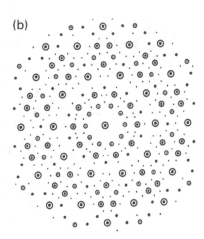

2.11. Electron diffraction pattern of an AlMn quasicrystal along the fivefold axis (left) and a computed Fourier pattern of a three-dimensional Penrose tiling (right). From C. Janot, *Quasicrystals, A Primer*, 2nd ed. (London: Oxford Univ. Press, 1994), p. 3, figure 1.24.

Fractals

Fractals are self-similar shapes that have similar appearances as they are magnified. This is called *dilational symmetry*. Each generation looks like the previous generation. Cauliflower is an example. Each branch looks just like a miniature of the whole head. Figure 2.12 shows an irregular fractal. Fractals that have greater symmetry are called *regular fractals*. Figure 2.13 is an example.

2.12. A two-dimensional projection of a three-dimensional irregular fractal.

A useful parameter is the fractal dimension, D, which is the exponent in the relation between the mass, M, to a linear dimension, R:

$$M = CR^D. \tag{2.1}$$

2.13. A regular fractal.

For example, in Figure 2.13, $M_2/M_1 = (R_2/R_1)^D$ so $D = \ln(M_2/M_1)/\ln(R_2/R_1)$. Here $M_2 = 54$, $M_1 = 9$, $R_2 = 9$, and $R_1 = 3$. Substituting these, $D = 1.63$. A solid object can be thought of as a fractal of $D = 3$.

Fractals find use in studies of fracture, surface roughness, and disordered materials.

NOTE OF INTEREST

M. C. Escher's woodcut *Heaven and Hell* (Figure 2.14) is an illustration of symmetry in art. It also is an example of fractals.

2.14. A woodcut titled *Heaven and Hell*, by M. C. Escher. From M. C. Escher, *The Graphic Work of M. C. Escher* (New York: Ballantine Books, 1967), plate 23.

SYMMETRY

REFERENCES

S. M. Allen and E. L. Thomas. *The Structure of Materials*. New York: Wiley, 1999.

B. D. Cullity and S. R. Stock. *Elements of X-Ray Diffraction*, 3rd ed. Englewood Cliffs, NJ: Prentice-Hall, 2001.

C. Janot. *Quasicrystals: A Primer*, 2nd ed. London: Oxford Univ. Press, 1994.

A. Kelly, G. W. Groves, and P. Kidd. *Crystallography and Crystal Defects*. New York: Wiley, 2000.

P. R. Massopust. *Chaos, Solitons and Fractals* 8 (1997): 171–90.

S. Ranganathan and K. Chattopadhyay. *Ann. Rev. Mater. Sci.* 21 (1991): 437.

PROBLEMS

1. Why is there no face-centered tetragonal space lattice? Why is there no base-centered tetragonal?
2. How many twofold axes of rotation are there in a simple hexagonal prism?
3. Deduce the five two-dimensional Bravais lattices.
4. Show that $\tau^2 - \tau - 1 = 0$, where τ is the golden ratio.
5. Calculate the packing factor for the first, second, and third generation of the fractal in Figure 2.12.
6. Calculate the fractal dimension for the two-dimensional fractal in Figure 2.15.

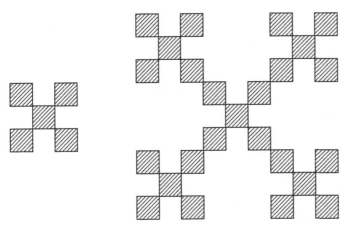

2.15. A two-dimensional fractal.

7. Describe the symmetry elements in Figure 2.14.

8. Describe the symmetry elements of a pentagonal dodecahedron. It has 12 faces and 30 edges. See Figure 2.16.

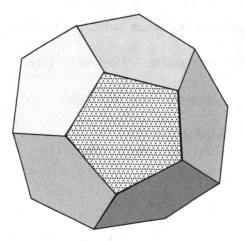

2.16. Pentagonal dodecahedron.

3 Miller–Bravais Indices for Hexagonal Crystals

The Miller–Bravais index system for identifying planes and directions in hexagonal crystals is similar to the Miller index system except that it uses four axes rather than three. The advantage of the four-index system is that the symmetry is more apparent. Three of the axes, a_1, a_2, and a_3, lie in the hexagonal (basal) plane at 120° to one another and the fourth or c-axis is perpendicular to then, as shown in Figure 3.1.

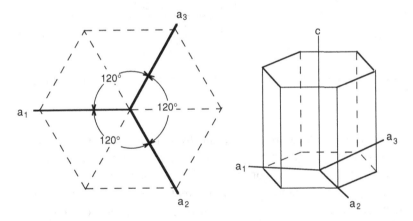

3.1. Axis system for hexagonal crystals.

Planar indices

The rules for determining Miller–Bravais planar indices are similar to those for Miller indices with three axes.

1. Write the intercepts of the plane on the four axes in order (a_1, a_2, a_3, and c).
2. Take the reciprocals of these.
3. Reduce to the lowest set of integers with the same ratios.
4. Enclose in parentheses $(hki\ell)$.

Commas are not used, except in the rare case that one of the integers is larger than one digit. (This is rare because we are normally interested only in directions with low indices.) If a plane is parallel to an axis, its intercept is taken as ∞ and its reciprocal as 0. If the plane contains one of the axes or the origin, either analyze a parallel plane or translate the axes before finding indices. This is permissible since all parallel planes have the same indices. Figure 3.2 shows several examples.

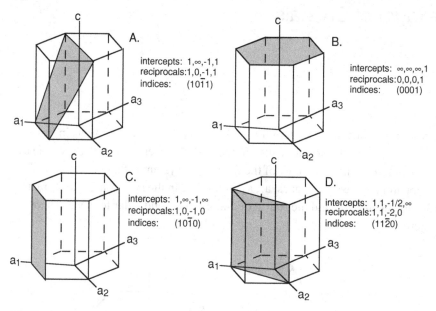

3.2. Examples of planar indices for hexagonal crystals. Note that the sum of the first three indices is always zero: $h + k + i = 0$.

In the four-digit system, the third digit, i, can always be deduced from the first two, $i = -h - k$, and is therefore redundant. With the three-digit systems, it may either be replaced by a dot, $(hk \cdot \ell)$, or omitted entirely, $(hk\ell)$. If the third index is omitted, the hexagonal symmetry is not apparent. In the four-digit Miller–Bravais system, families of planes are apparent from the indices. For example, $\{01\bar{1}0\} = (01\bar{1}0), (\bar{1}010),$ and $(1\bar{1}00)$. The equivalence of the same family is not so apparent in the three-digit system, $\{010\} = (010)\{010\} = (010), (\bar{1}00),$ and $(\bar{1}10)$. Also compare $\{\bar{2}110\} = (\bar{2}110), (1\bar{2}10),$ and $(11\bar{2}0)$ with $\{\bar{2}10\} = (\bar{2}10), (1\bar{2}0),$ and (110). $\{\bar{2}10\} = (\bar{2}10), (1\bar{2}0)$.

Direction indices

The direction indices are the translations parallel to the four axes that produce the direction under consideration. The first three indices must be chosen so that they sum to zero and are the smallest set of integers that will express the direction. For example, the direction parallel to the a_1 axis is $[2\bar{1}\bar{1}0]$. The indices are enclosed without commas in brackets $[hki\ell]$. Examples are shown in Figure 3.3.

MILLER–BRAVAIS INDICES FOR HEXAGONAL CRYSTALS

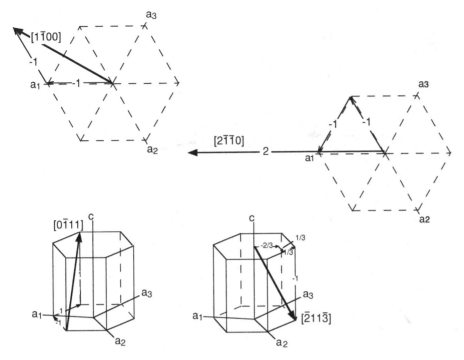

3.3. Examples of direction indices with the Miller–Bravais system.

Three-digit system

There is also the three-digit system for directions in hexagonal crystals. For planar indices, it uses intercepts on the a_1, a_2, and c axes. The indices (HKL) are related to the Miller–Bravais indices $(hki\ell)$ by

$$H = 2h + k + \ell, K = k - h + \ell, L = -2k - h + \ell, \qquad (3.1)$$
$$h = (1/3)(H - K), k = (1/3)(K - L), i = -(h + k), \quad \text{and}$$
$$\ell = (1/3)(H + K + L). \qquad (3.2)$$

The direction indices use the translations along the a_1, a_2, and c axes (U, V, and W, respectively). The four-digit $[uvtw]$ and three-digit $[UVW]$ systems are related by

$$U = u - t, \quad V = v - t, \quad W = w, \qquad (3.3)$$

and

$$u = (2U - V)/3, v = -(2V - U)/3,$$
$$t = -(u + v) = -(U + V)/3, w = W. \qquad (3.4)$$

The four- and three-digit systems are compared in Figure 3.4 for four directions.

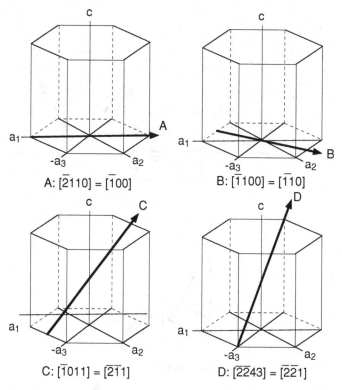

3.4. Comparison of the four- and three-digit systems.

NOTE OF INTEREST

Auguste Bravais (1811–1863) first proposed the Miller–Bravais system for indices. Also, as a result of his analyses of the external forms of crystals, he proposed the 14 possible space lattices in 1848. His *Études Cristallographiques*, published in 1866, after his death, treated the geometry of molecular polyhedra.

REFERENCES

C. S. Barrett and T. B. Massalski. *Structure of Metals*, 3rd ed. New York: McGraw-Hill, 1980.

A. Kelly, G. W. Groves, and P. Kidd. *Crystallography and Crystal Defects*. New York: Wiley, 2000.

PROBLEMS

1. Write the correct direction indices, [ ], and planar indices, ( ), for the directions and planes in Figure 3.5.

MILLER–BRAVAIS INDICES FOR HEXAGONAL CRYSTALS

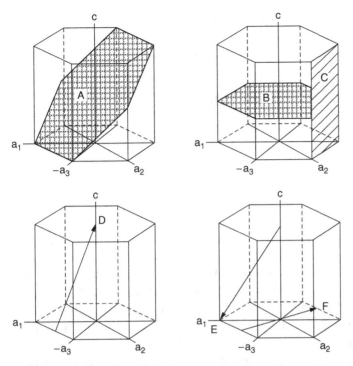

3.5. Several planes and directions for Problem 1.

2. Translate (210) in the three-digit system into the four-digit Miller–Bravais system. What are the equivalent {210} planes? Express these in both the three-digit and four-digit systems.

3. Sketch the ($\bar{2}$112) plane in a hexagonal cell.

4 Stereographic Projection

Projection

Stereographic projection provides a convenient way of displaying the angular relations between planes and directions in a crystal in two dimensions. The system involves first projecting planes and directions of interest onto a spherical surface and then mapping the spherical surface. Figure 4.1 illustrates how planes and directions are projected onto a sphere. If an infinitesimal crystal were placed at the center of a sphere and its planes extended, they would intersect the sphere as great circles and their directions would intersect the sphere as points.

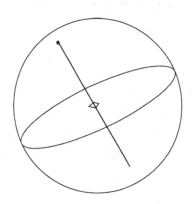

4.1. Mapping of planes and directions by placing an infinitesimal crystal at the center of a sphere and projecting planes onto the sphere to form great circles and lines to form points.

Projecting these points and great circles onto a flat surface is the same problem as projecting the earth's surface to form maps. The stereographic projection can be envisioned as placing the reference sphere on a plane and having a light source on the surface opposite the point of tangency. The light source then projects the lower half of the sphere onto the flat surface, as shown in Figure 4.2. Because of the symmetry of crystals, only one hemisphere need be mapped.

STEREOGRAPHIC PROJECTION

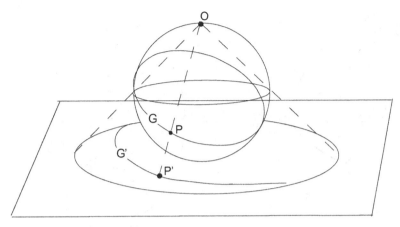

4.2. With stereographic projection, the elements on the lower half of the sphere are projected onto a flat surface from a point opposite the point of tangency.

Standard cubic projection

Of particular interest is the standard projection of a cubic crystal (Figure 4.3). The [001] direction is at the north pole, so the (001) plane forms the equator. The [100] direction is at the center, and the (100) plane forms the reference circle. The [010] direction is on the equator and the reference circle, and the (010) plane is a vertical line through the center. The <100> directions are represented by squares to symbolize their fourfold symmetry.

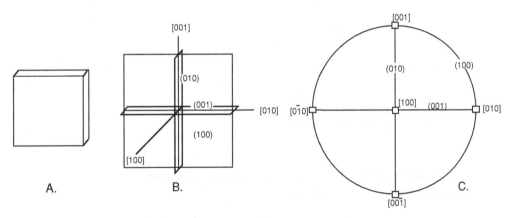

4.3. Projection of cube planes and directions.

The [0$\bar{1}$1] and [011] directions (and their opposite ends, [01$\bar{1}$] and [0$\bar{1}$$\bar{1}$]) are indicated by ellipses because they have twofold symmetry (Figure 4.4A). They are on the reference circle, 45° from ±[010] and ±[001]. The corresponding (0$\bar{1}$1) and (011) planes cut diagonally through the center. Note that (0$\bar{1}$1), [100], and (01$\bar{1}$) all lie on the (011) plane because they are 90° from [011]. The (110), (1$\bar{1}$0), (101), and (10$\bar{1}$) planes and their normals are similarly constructed and symbolized.

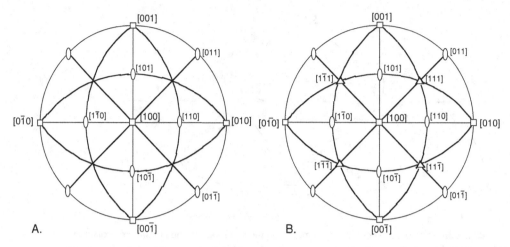

4.4. Construction of the standard cubic projection.

There are four points at which three great circles representing the {110} intersect. These points must be the directions that lie in all three of those planes, namely, the <111> directions. Triangles are used as symbols for the <111> directions because they have threefold symmetry (Figure 4.4B). This construction divides orientation space into 24 spherical triangles, each of which have <100>, <110>, and <111> corners (Figure 4.5). All of the triangles are crystallographically equivalent.

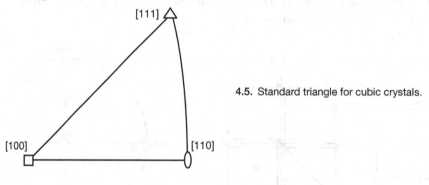

4.5. Standard triangle for cubic crystals.

Locating the *hkℓ* pole in the standard stereographic projection of a cubic crystal

Consider the standard projection with [100] at the center and [001] at the north pole (Figure 4.6). For all poles in the projected hemisphere the index, h, is positive because these poles lie 90° or less from [100] at the center, so their dot product $hk\ell$ with [100] is positive. Poles on the outer circumference are 90° from [100], so for these $h = 0$. Similarly, the projected hemisphere can be divided into four quadrants. In the first quadrant $k > 0$ and $\ell > 0$ because all poles in this quadrant are less than 90° from both [010] and [001]. In the second quadrant $k < 0$ because poles in this region are more than 90° from [010]. Both k and ℓ are negative in the third quadrant because poles in this region are more than 90° from both [010] and [001]. Finally, $\ell > 0$ in the fourth quadrant because poles in this region are 90° or more from [001].

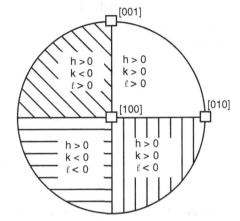

4.6. Signs of h, k, and ℓ in the four quadrants of the standard projection.

Similarly, each quadrant can be divided in half according to whether $h > k$, whether $k > \ell$, and whether $h > \ell$, as shown in Figure 4.7 for the first quadrant. These three bisections of the first quadrant split it into six triangles, as illustrated in Figure 4.8 with the appropriate <123> pole in each triangle.

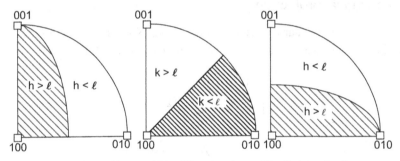

4.7. Relative values of h, ℓ, and k in different regions of the first quadrant.

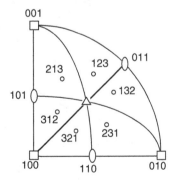

4.8. Locations of the <123> poles in the first quadrant.

EXAMPLE 4.1. Locate $[2\bar{1}1]$ on the standard projection.

SOLUTION: See Figure 4.9. Examination of the signs of each of the indices shows that $[2\bar{1}1]$ lies in the fourth quadrant (Figure 4.9A). The dot products $[2\bar{1}1] \cdot [0\bar{1}0]$

and $[2\bar{1}1] \cdot [001]$ are equal, so $[2\bar{1}1]$ is equidistant from $[0\bar{1}0]$ and $[001]$ on the line connecting $[100]$ and $[0\bar{1}1]$ (Figure 4.9B). Finally, the dot product of $[2\bar{1}1]$ with $[100]$ is larger than the dot products of $[2\bar{1}1]$ with $[0\bar{1}1]$. This indicates that the angle between $[2\bar{1}1]$ and $[100]$ is less than the angles between $[2\bar{1}1]$ and $[0\bar{1}1]$ (Figure 4.9C).

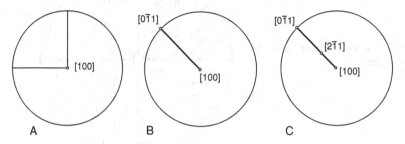

4.9. Location of $[2\bar{1}1]$ on the standard projection.

Other projections are possible, but crystallographers use the stereographic projection because the angles between planes in the projection are the true angles between the planes.

Standard hexagonal projection

Figure 4.10 shows the standard hexagonal projection. Note that there are 12 equivalent triangles with corners at $<2\bar{1}\bar{1}0>$, $<10\bar{1}0>$, and $[0001]$.

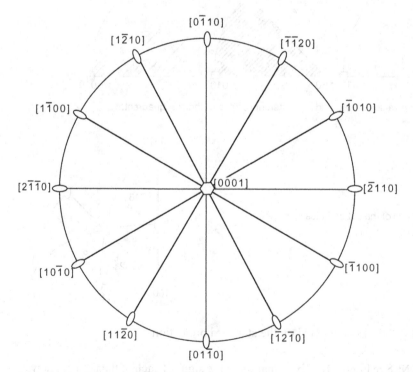

4.10. Standard hexagonal projection.

STEREOGRAPHIC PROJECTION

Spherical trigonometry

The angular relations expressed graphically by the stereographic projection can be expressed quantitatively by spherical geometry. A spherical triangle is a triangle on the surface of a sphere whose sides are great circles, as shown in Figure 4.11. Two very simple and useful relations are

$$\cos a = \cos b \cos c + \sin b \sin c \cos \alpha \qquad (4.1)$$

and

$$\cos \alpha = \cos \beta \cos \gamma + \sin \beta \sin \gamma \cos a, \qquad (4.2)$$

where a, b, and c are the angular lengths of the sides and α, β, and γ are the interior angles.

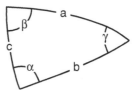

4.11. Spherical triangle.

NOTE OF INTEREST

Of the various ways of projecting a spherical surface onto a plane, the stereographic projection is the only one that preserves angles between directions and preserves circles as circles. The word "stereographic" comes from the Greek *stereos* meaning solid and *graphicus* meaning writing, drawing, or engraving.

REFERENCES

C. S. Barrett and T. B. Massalski. *Structure of Metals*, 3rd ed. New York: McGraw-Hill, 1980.
W. F. Hosford. *The Mechanics of Crystals and Textured Polycrystals*. Oxford, U.K.: Oxford Science, 1993.
A. Kelly, G. W. Groves, and P. Kidd. *Crystallography and Crystal Defects*. New York: Wiley 2000.

PROBLEMS

1. Locate [312] on a stereographic projection with [001] at the north pole and [100] in the center.
2. What is the angle between the [110] and [321] directions in a cubic crystal?

3. How many different <210> directions are there in a cubic crystal? How many stereographic triangles does each <210> direction share?

4. Calculate the great circle distance from Chicago (lat. = 41.8 N, long. = 87.75 W) and Auckland (lat. = 36.75 S, long. = 174.75 E). The earth's diameter is 12,742 km.

5 Crystal Defects

Vacancies in pure metals

A common lattice defect is a vacant lattice site or vacancy. The presence of a vacancy increases the enthalpy by Δh_f and the entropy by Δs_f. Because the free energy of a system is lowered by increased entropy, there is an equilibrium fraction of vacant lattice, x_v, which increases with temperature. At equilibrium, $\Delta g_f = \Delta h_f - T\Delta s_f = 0$, so $\Delta h_f = -T\Delta s_f$. From statistical mechanics, $\Delta s_f = -k \ln x_v$, so

$$x_v \approx \exp(-\Delta h_f/kT), \tag{5.1}$$

where k is Boltzmann's constant, 86.1×10^{-6} eV/K. Data for pure metals (Figure 5.1) indicate that $\Delta h_f \approx 0.75 \times 10^{-3} T_m$, where T_m is the melting point in Kelvins. This means that at the melting point of a metal, the equilibrium concentration, x_v, is about 2×10^{-4} and at $T_m/2$, $x_v \approx 5 \times 10^{-7}$.

Actual vacancy concentrations at low temperatures are often much higher than the equilibrium number. There are three reasons for this: First, as a metal is cooled the number of vacancies can decrease only by diffusing to sinks, such as edge dislocations, grain boundaries, or free surfaces. Unless cooling from a high temperature is very slow, there will not be enough time for all of the excess vacancies to diffuse to these sinks. A second reason is that, during plastic deformation, intersecting dislocations create jogs, and the movement of jogged dislocations generates vacancies. Radiation is a third possible cause of excess vacancies. Neutrons may knock atoms out of their normal lattice positions, creating both vacancies and interstitial atoms. This is particularly important in nuclear reactors.

The increase of resistivity after cold work is caused by vacancies. Recovery annealing decreases vacancy concentration, restoring conductivity.

In principle, there should also be an equilibrium concentration of interstitial atoms that increases with temperature:

$$x_i \approx \exp(-\Delta h_i/kT), \tag{5.2}$$

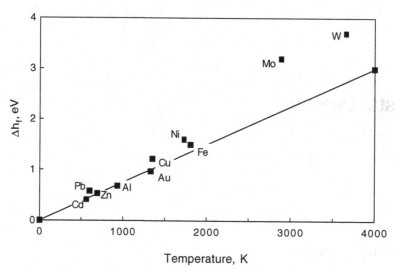

5.1. Correlation of the heat of formation of vacancies with melting points. Data from R. A. Johnson in *Diffusion* (Materials Park: ASM, 1972) All rights reserved. www.asminternational.org.

where Δh_i is the energy to create an interstitial. However, estimated values of Δh_i are so high that even at the melting point x_i is negligible.

Interstitials may be generated by radiation.

Point defects in ionic crystals

Point defects in ionic crystals are always paired to preserve electrical neutrality. There are several types of defect pairs:

1. A cation vacancy may be paired with an anion vacancy. This is called a *Schottky defect*. An example is the formation of Li^+ and F^- vacancies in LiF. This is illustrated in Figure 5.2A.
2. A cation vacancy may be paired with a nearby cation interstitial. This is called a *Frenkel pair*. An example is the formation of Zn^{+2} vacancies and Zn^{+2} interstitials in ZnO. This is illustrated in Figure 5.2B. In principle, paired anion vacancies and interstitials are possible, but this is less likely because of the larger size of the anions.
3. A solution of cations of a higher valence than that of the solvent cation must be accompanied by one or more anion vacancies. For example, consider the presence of Fe^{+3} ions in FeO. For every two Fe^{+3} ions, there is an O^{-2}, vacancy, as shown in Figure 5.2C.

There is a useful notation system for defects in ionic crystals. Either the chemical symbol for the element or "V" for a vacancy indicates what is on a lattice site. A subscript of either the chemical symbol for the element or "i" if it is an interstial site indicates what is normally on the site. The charge, relative to the

CRYSTAL DEFECTS

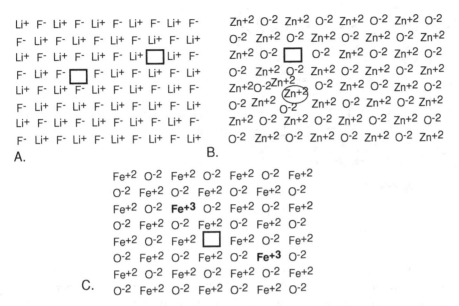

5.2. (A) Shottky defects consist of paired cation and anion vacancies. (B) Frenkel defects consist of ion vacancy and ion interstitial pairs. (C) Anion vacancies may be neutralized by substitution of cations of higher valence.

normal charge of the site, is indicated by a superscript + for positive charges, − for negative charges or ° for no change of charge.

For example, using this system, V^-_{Na} indicates a vacant site normally occupied by a Na$^+$ ion so the site has a charge of −1, Fe^+_{Fe} indicates an Fe^{+++} substituting for an Fe^{++} so the site has a charge of +1, and $Ag°_{Ag}$ indicates a silver ion on its proper site. Reactions can be written with this system. For example, formation of a Frenkel defect in NaCl can be written as $Na°_{Na} \rightarrow Na^+_i + V^-_{Na}$ and the creation of a Schottky defect can be written Null $\rightarrow V^-_{Na} + V^+_{Cl}$.

The enthalpies associated with several reactions are listed in Table 5.1 The equilibrium number of defect pairs is given by

$$n = \exp(-\Delta g/2kT) \approx \exp(-\Delta h/2kT). \quad (5.3)$$

Substitution of an ion having a different valence than the solute causes a defect. For example, solution of CaCl$_2$ in NaCl creates a vacancy on a Na$^+$ site. The

Table 5.1. Formation enthalpies for vacancy formation in a few ionic crystals

Crystal	Reaction	Formation enthalpy, Δh, eV
AgBr	$Ag°_{Na} \rightarrow Ag^+_i + V^-_{Ag}$	1.1
LiF	$Li°_{Li} \rightarrow Li^+_i + V^-_{Li}$	2.4 – 2.7
TiO$_2$	$O°_O \rightarrow V^{2+}_O + O^{2-}_i$	8.7
ZnO	$O°_O \rightarrow V^{2+}_O + O^{2-}_i$	2.5
Al$_2$O$_3$	Null $\rightarrow 2V^{3+}_{Al} + 3^{2+}°_O$	26
FeO	Null $\rightarrow Fe^{2-}_{Fe} + V^{2+}_O$	6.5
MgO	Null $\rightarrow Mg^{2-}_{Mg} + V^{2+}_O$	7.7

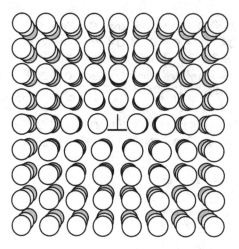

5.3. An edge dislocation in a simple cubic crystal.

solution reaction can be written as $CaCl_2 \rightarrow Ca^+{}_{Na} + V^-{}_{Na} + 2Cl^\circ{}_{Cl}$. Defects caused by impurities are called *extrinsic* defects in contrast to temperature-dependent *intrinsic* defects. In ionic crystals, point defects can act as charge carriers. Therefore, the electrical conductivity increases with an increasing number of defects.

Dislocations

A dislocation is a line defect in a crystal. The atoms around a dislocation are displaced from their normal positions. The lattice distortion is greatest near the dislocation and decreases with distance from it. Figure 5.3 shows an edge dislocation, which is a special form of dislocation. Its geometry is equivalent to cutting into a perfect crystal and inserting an extra half plane of atoms into the cut, as sketched in Figure 5.4. The dislocation is the bottom edge of the extra half plane.

A more general way of visualizing a dislocation is to imagine cuting into a crystal and shearing one side of the cut relative to the other by an atomic distance. This is illustrated in Figure 5.5. If the direction of shearing is perpendicular to the end of the cut, the end of the cut is an edge dislocation. If the shearing is parallel to the end of the cut, the end of the cut is a screw dislocation. The atoms around

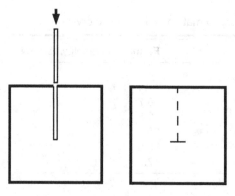

5.4. Creation of an edge dislocation by insertion of an extra half plane of atoms.

CRYSTAL DEFECTS

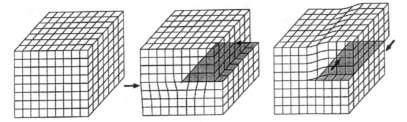

5.5. Creation of an edge dislocation by shearing perpendicular to a cut (middle) and a screw dislocation by shearing parallel to a cut (right.) From W. F. Hosford, *Physical Metallurgy* (Cleveland, OH: CRC Press, 2004), p. 126, figure 8.2.

it are on planes that spiral around the dislocation like the ramps of a parking structure.

In both cases the dislocation is the boundary between the region that has been sheared and the region that has not. The boundary need not be either perpendicular or parallel to the direction of slip. Edge and screw orientations are extremes. A dislocation need not be a straight line. However, as a dislocation wanders through a crystal, its Burgers vector is always the same. When a dislocation moves, it causes some material to undergo slip.

Burgers vectors

A dislocation is characterized by its *Burgers vector*. An atom-to-atom circuit that would close in a perfect crystal will fail to close if it is drawn around a dislocation. The closure failure is the Burgers vector of the dislocation. This is illustrated in Figure 5.6. The edge dislocation (middle) is perpendicular to its Burgers vector and the screw dislocation (right) is parallel to its Burgers vector.

A Burgers vector has both a direction and a magnitude. Its direction is the direction of the displacement that would be caused by movement of the dislocation, and the magnitude is the length of that displacement. The direction and magnitude normally correspond to a slip direction and slip displacement. The common notation indicates the direction by Miller indices. A scalar in front indicates the magnitude. For example, $\mathbf{b} = (a/2)[110]$ indicates a vector $a/2, a/2, 0$, where

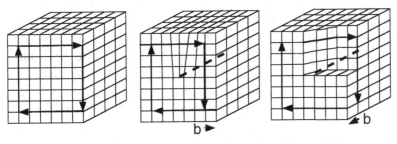

5.6. A circuit that closes in a perfect crystal (left). If that circuit is drawn around an edge dislocation (dashed line), the closure failure, **b**, is perpendicular to the dislocation (middle). Around a screw dislocation, the closure failure, **b**, is parallel to the dislocation (right).

a is the lattice parameter. This is the Burgers vector corresponding to a full slip displacement in an fcc crystal. Its magnitude is $(a/2)[1^2 + 1^2 + 0^2]^{1/2} = a/\sqrt{2}$.

Energy of dislocations

The energy per length of a screw dislocation is approximately

$$E_L = Gb^2, \tag{5.4}$$

where G is the shear modulus of the crystal. For an edge dislocation the energy is about one and a half times greater, $E_L = Gb^2/(1 - \nu)$, where ν is Poisson's ratio. These expressions of energy per length are equivalent to line tensions. Dislocations tend to straighten themselves to minimize their energy.

Two parallel dislocations can combine to form a third dislocation, or one dislocation can dissociate into two others. In either case, the vector sums of the products and reactants must be equal. That is, if $b_1 + b_2 \to b_3 + b_4$, $b_1 + b_2 = b_3 + b_4$. Since the energy of a dislocation is proportional to b^2, dislocation reactions are energetically favorable if $b_1^2 + b_2^2 > b_3^2 + b_4^2$ and unfavorable if $b_1^2 + b_2^2 < b_3^2 + b_4^2$. This is known as Frank's rule. For example, the reaction $(a/2)[110] + (a/2)[011] \to (a/2)[121]$ is unfavorable because $(a/2)^2[1^2 + 1^2 + 0^2] + (a/2)^2[0^2 + 1^2 + 1^2] = a^2$ is less than $(a/2)^2[1^2 + 2^2 + 1^2] = (3/2)a^2$.

Stress fields around dislocations

The displacements of atoms near a dislocation from their normal lattice positions are the same as the displacements that would be caused by some external stress. Therefore, we can think of the dislocation as causing a stress field around it. Around an edge dislocation, there is a state of hydrostatic stress, $\sigma_H = \sigma_x + \sigma_y + \sigma_z$, at a location x, y,

$$\sigma_H = Ay/(x^2 + y^2), \tag{5.5}$$

where $A = Gb(1 + \nu)/[3\pi(1 - \nu)]$. This is illustrated in Figure 5.7. Above the dislocation the extra half plane causes crowding. This is equivalent to hydrostatic compression. There is hydrostatic tension below the dislocation. The level of the hydrostatic stress decreases with distance from the dislocation.

The stress field around an edge dislocation causes interactions with solutes. Interstitial solutes tend to segregate to the region of hydrostatic tension. Substitutional solutes that are larger than the lattice atoms also segregate to the region of the hydrostatic tension. Small substitutional solutes will segregate to the region of compression. In all cases the segregation lowers the energy of the dislocation, making it more difficult to move. There is no dilation around a screw dislocation in an isotropic crystal so there is little interaction between solutes and screw dislocations.

The stress fields of edge dislocations interact with other edge dislocations. The systems' energy is lowered if they are aligned so that the compressive field of one

CRYSTAL DEFECTS

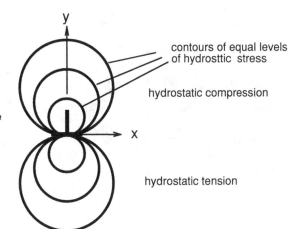

5.7. Hydrostatic stress field around an edge dislocation.

overlaps the tensile field of another. Such an arrangement forms a low-angle grain boundary, as shown in Figure 5.8. The angle of misorientation, θ, is given by

$$\theta = b/d. \tag{5.6}$$

5.8. Low–angle grain boundary formed by a series of edge dislocations.

Partial dislocations

In fcc metals, the normal slip dislocations can dissociate into partial dislocations:

$$(a/2)[110] \rightarrow (a/6)[121] + (a/6)[21\bar{1}]. \tag{5.7}$$

This reaction is energetically favorable. The resulting $(a/6) < 211 >$ dislocations are called partial dislocations because their movement does not restore the lattice. They repel each other but leave a stacking fault between them. This is illustrated in Figure 5.9. The stacking of close-packed planes in this region corresponds to a three-atom layer of hexagonal close-packed (hcp) packing (Figure 5.10). This is not the normal stacking and, therefore, there is a stacking fault energy associated with such a region.

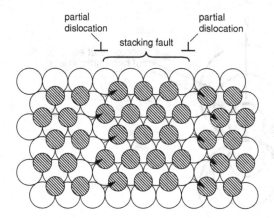

5.9. Between two partial dislocations there is a stacking fault. Reproduced with permission of Cambridge University Press from W. F. Hosford, *Mechanical Behavior of Materials* (New York: Cambridge Univ. Press, 2005).

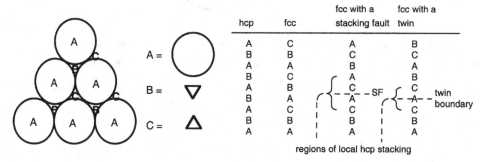

5.10. A stacking fault corresponds to a thin region of hcp packing. Reproduced with permission of Cambridge University Press from W. F. Hosford, *Mechanical Behavior of Materials* (New York: Cambridge Univ. Press, 2005).

NOTES OF INTEREST

1. The term *ion* was first used in 1834 by Michael Faraday. It comes from a Greek word meaning "to go." He also named *anion* (thing going up) and *cation* (thing going down.)
2. Early theoretical chemists studying how crystals grow realized that once an atomic plane was complete another would have to be nucleated. However, their calculations indicated that nucleation of new planes would be far too slow to account for observed rates of crystal growth. The postulation of screw dislocations removed this dilemma because nucleation of new planes would not be required.
3. As part of an investigation, in 1952, of the failure of some electrical condensers, Treuting found whiskers of tin (about 2×10^{-6} m diameter) growing from the condenser walls, as shown in Figure 5.11. When these whiskers were tested in bending,* $(a/2)[110] \rightarrow (a/6)[2\bar{1}1] + (a/6)[121]$, it was found that these whiskers could be bent to a strain of 2 to 3% without plastic deformation. This meant that the yield strength was more than 2.5% of

---

* C. Herring and J. K. Galt, *Phys. Rev.* 85 (1952): 1060–1.

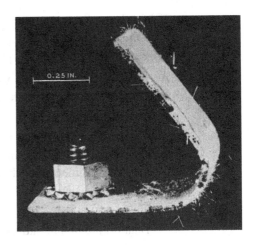

5.11. Tin whiskers growing on tin-plated steel. From W. C. Ellis, D. F. Gibbons, and R. G. Treuting in *Growth and Perfection of Crystals*, B. H. Doremus, B. W. Roberts, and D. Turnbull, eds. (New York: Wiley, 1958), p.102.

the Young's modulus. For tin, $E = 44$ GPa, so the yield strength was over 1 GPa. Overnight, tin became the strongest material known to man. Once the Herring–Galt observation was reported, many others began testing whiskers of other metals such as copper and iron with similar results and tin lost this distinction.

REFERENCES

F. Agullo-Lopez, C. R. A. Catlow, and P. D. Townsend. *Point Defects in Materials*. New York: Academic Press, 1988.

A. H. Cottrell. *Dislocations and Plastic Flow in Crystals*. London: Oxford Univ. Press, 1953.

J. I. Gersten and F. W. Smith. *The Physics and Chemistry of Materials*. New York: Wiley, 2001.

J. P. Hirth and J. Lothe. *Theory of Dislocations*. New York: Wiley, 1982.

W. F. Hosford. *Mechanical Behavior of Materials*. New York: Cambridge Univ. Press, 2005.

D. Hull and D. J. Bacon. *Introduction to Dislocations*, 3rd ed. London: Butterworth Heinemann, 1997.

F. A. Kroger. *The Chemistry of Imperfect Crystals*. Amsterdam: North Holland, 1964.

W. T. Read. *Dislocations in Crystals*. New York: McGraw-Hill, 1953.

J. Weertman and J. R. Weertman. *Elementary Dislocation Theory*. London: Oxford Univ. Press, 1992.

PROBLEMS

1. The composition of wustite is $Fe_{<1}O$. The equilibrium ratio Fe/O depends on the oxygen partial pressure. The primary charge carriers are electron holes (Fe^{+3} ions). Explain how the conductivity of wustite changes with oxygen partial pressure.

2. It is estimated that the energy to form an interstitial atom in a metal is four times as great as that to form a vacancy. Estimate the equilibrium concentration of interstitials just below the melting point.

3. A. Estimate the equilibrium vacancy concentration of aluminum at 25 °C.
   B. Give two reasons why the actual concentration might be much higher than this amount.

4. A crystal of aluminum contains 10^{12} meters of dislocation per cubic meter.
   A. Calculate the total amount of energy associated with dislocations 10^{12} per m^3. Assume half of the dislocations are edges and half are screws.
   B. If all of this energy could be released as heat, what would be the temperature rise?

   Data for aluminum: atomic diameter = 0.286 nm,
   crystal structure = fcc, density = 2.70 Mg/m^3,
   atomic mass = 27 g/mol, $C = 0.90$ J/g C, $G = 70$ GPa, $\nu = 0.3$.

5. Calculate the average spacing between dislocations in a 1/2° tilt boundary in aluminum. Look up any required data.

6. Consider the reactions between parallel dislocations given below. In each case write the Burgers vector of the product dislocation and determine whether the reaction is energetically favorable.
   A. $(a/2)(a/2)[1\bar{1}0] + (a/2)[110] \to$
   B. $(a/2)[101] + (a/2)[01\bar{1}] \to$
   C. $(a/2)[1\bar{1}0] + (a/2)[101] \to$

7. Consider the dislocation dissociation reaction $(a/2)[110] \to (a/6)[2\bar{1}1] + (a/6)[121](a/2)[1\bar{1}0] + (a/2)[101] \to$ in an fcc crystal. Assume that the energy/length of a dislocation is given by $E_L = Gb^2$ and neglect any dependence of the energy on the edge versus the screw nature of the dislocation. Assume that this reaction occurs and the partial dislocations move very far apart.
   A. Express the total decrease in energy/length of the original $(a/2)[110]$ dislocation in terms of a and G.
   B. On which $\{111\}$ plane must these dislocations lie?

8. Referring to Figure 5.10, find the ratio of the wrong second-nearest neighbors across a stacking fault to the number across a twin boundary. If the surface energies are proportional to the number of wrong second-nearest neighbors, what is γ_{SF}/γ_{TB}?

6 Phase Diagrams

The Gibbs phase rule

The Gibbs phase rule relates the number of equations that are required to describe a system at equilibrium to the number of variables necessary to describe the system. The number of degrees of freedom is the number of variables that can be changed independently without affecting the number of phases in equilibrium. It is the difference between the number of variables and the number of equations describing equilibrium.

Variables: The variables in the system are the composition of *each phase* and the environmental variables. $C-1$ independent terms are needed to express the composition of each phase, where C is the number of components (elements or compounds). For example, to describe the composition of the α phase we would need to fix $c_A^\alpha, c_B^\alpha, \ldots, c_C^\alpha - 1$, where c_A^α is the amount of A in the α phase. The reason that the number of compositional variables for the α phase is $C-1$ rather than C is that once the percent (or fraction) of all but one of the components has been established, the amount of the last one is fixed. With P phases and $C-1$ compositional variables for each phase, the total number of compositional variables is $P(C-1)$.

The usual environmental variables are temperature and pressure. However, one can imagine a system in which equilibrium is affected by some other variable (e.g., a magnetic field). Pressure is not considered to be a variable when equilibrium is described at a fixed total pressure (e.g., atmospheric). In general the number of environmental variables is designated as E. Most phase diagrams are at constant pressure so the only environmental variable is temperature. In this case $E = 1$. (If both variations in both temperature and pressure are considered, $E = 2$).

Equalities: In a phase at equilibrium there is no driving force for a change of composition. Therefore, the vapor pressure, p, of each element must be the same in all phases.* For example, if the equilibrium vapor pressure of water vapor were

---

* Some authors prefer to describe the equilibrium in terms of partial free energies or chemical activities instead of vapor pressures. Because these quantities are uniquely related to vapor pressures, the result is the same.

higher for the α phase than for the β phase, water vapor would evaporate at the α, and diffuse to and condense at the β. It does not matter if the vapor pressures are extremely low. Equilibrium therefore implies that

$$p_A{}^\alpha = p_A{}^\beta = p_A{}^\gamma = \ldots$$
$$p_B{}^\alpha = p_B{}^\beta = p_B{}^\gamma = \ldots \qquad (6.1)$$
$$p_C{}^\alpha = p_C{}^\beta = p_C{}^\gamma = \ldots$$

The subscripts refer to the components and the superscripts refer to the phases. For example, $p_A{}^\beta$ is the equilibrium vapor pressure of element A in phase β. With P phases (α, β, γ, ...) there are $P-1$ equalities (equal signs) in each line. There are C lines, where C is the number of components (A, B, C, etc.) so the total number of equalities is $C(P-1)$.

Degrees of freedom: The number of degrees of freedom, F, is the number of variables that can be changed independently without changing which phases are in equilibrium. It is the number of variables that are not fixed by the equalities, so it equals the number of variables, $P(C-1) + E$, minus the number of equalities, $C(P-1)$,

$$F = P(C-1) + E - C(P-1) \quad \text{or}$$
$$F = C + E - P. \qquad (6.2)$$

This is the Gibbs phase rule. For constant pressure but variable temperature it can be written as $F = C - P + 1$.

One simple way of remembering the phase rule is to write it as $P + F = C + E$, which might also be a shorthand way of saying a Police Force equals Cops plus Executives.)

Invariant reactions

Invariant reactions are ones that occur at constant temperatures. In a binary system, three phases may be in equilibrium at a constant temperature. A number of common invariant reactions are illustrated in Figure 6.1.

Ternary phase diagrams

In systems involving three components, composition is plotted on a triangular section (Figure 6.2). Pure components are represented at the corners and the grid lines show the amount of each component. All of the lines parallel to AB are lines on which the %C is constant. Those nearest C have the greatest amount of C. To represent temperature, a third dimension is needed. Figure 6.3 is a sketch of a three-dimensional ternary diagram in which temperature is the vertical coordinate.

Ternary equilibrium can be represented in two dimensions by either isothermal or vertical sections. A typical isothermal section is indicated in Figure 6.4. There are single-phase regions (α, β, γ, and δ), two-phase regions

PHASE DIAGRAMS

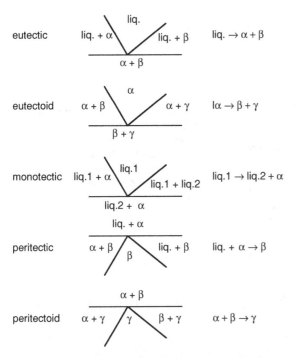

6.1. Several invariant reactions.

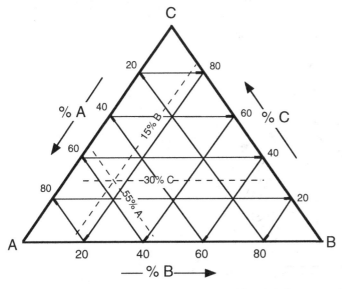

6.2. A triangular grid for representing the compositions in a three-component system.

($\alpha + \gamma$, $\alpha + \delta$, $\delta + \gamma$, $\delta + \beta$, and $\beta + \gamma$), and three-phase regions ($\alpha + \delta + \gamma$ and $\delta + \beta + \gamma$). As in binary diagrams, there are always two-phase regions between single-phase regions. Three-phase regions are triangular and meet single-phase regions only at their corners.

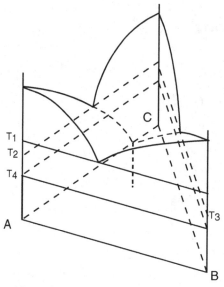

6.3. A three-dimensional model of a ternary phase diagram with a ternary eutectic. T_1 corresponds to the AB binary eutectic temperature, T_2 to the AC binary eutectic temperature, T_3 to the BC binary eutectic temperature, and T_4 to the ABC ternary eutectic temperature.

Two-phase regions: In two-phase regions, tie lines are needed to indicate the compositions of the two phases that are in equilibrium with each other. Often the tie lines are not shown. However, they can usually be approximated with a little judgment. The lever law may be used to find the relative amounts of the two phases at opposite ends of a tie line going through the overall composition. (In applying the lever law, the composition may be expressed in terms of any of the three components but greatest accuracy will be obtained by using the component that differs the most between the two phases.)

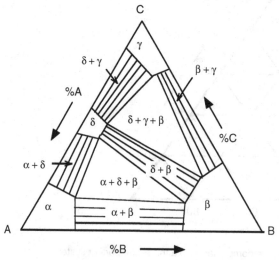

6.4. An isothermal section of a hypothetical ternary phase diagram showing one-, two-, and three-phase regions.

Three-phase regions: These are triangular. The compositions of the phases are at the corners. The relative amounts of the three phases can be found from a modified lever law,

$$f_\alpha = (C_{av} - C_\alpha)/(C_{\alpha'} - C_\alpha), \qquad (6.3)$$

PHASE DIAGRAMS

where C_α' is found by extrapolating a line drawn through C_α and C_{av} to the opposite side of the triangle, as shown in Figure 6.5.

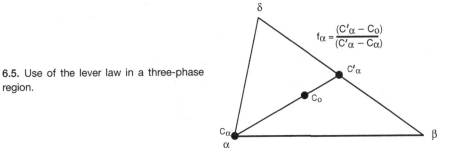

6.5. Use of the lever law in a three-phase region.

The progression of phases during solidification can be followed from a projection of the liquidus on triangular coordinates. Figure 6.6 shows the liquidus surface of the $CaO - Al_2O_3 - SiO_2$ system. During freezing the composition of the liquid moves away from the solid phase that is forming and down the temperature

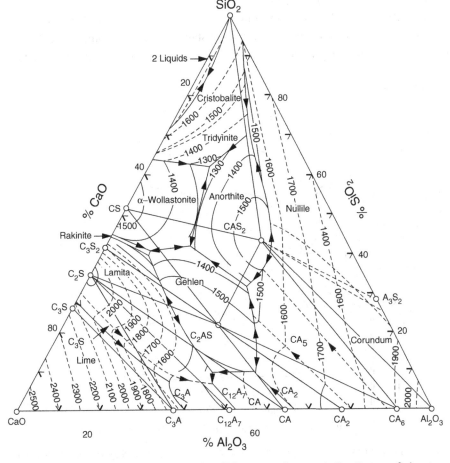

6.6. The liquidus surface of the $CaO-Al_2O_3-SiO_2$ ternary diagram. In the diagram, C denotes CaO, A denotes Al_2O_3, and S denotes SiO_2. Temperatures in Celsius. From *Phase Diagrams for Ceramic* 1 (1964): 219.

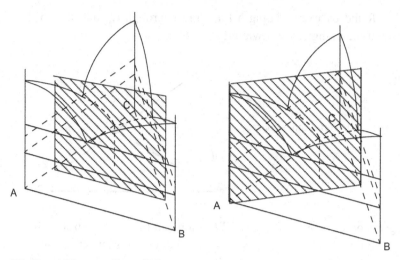

6.7. Pseudobinary sections of a ternary.

gradient. Once a eutectic valley is reached, the liquid composition follows the eutectic valley toward lower temperatures.

Vertical sections: Vertical cuts through a ternary can be made at either a constant amount of one of the components or at a constant ratio of two components (Figure 6.7). These pseudobinaries differ from binary phase diagrams in that the tie

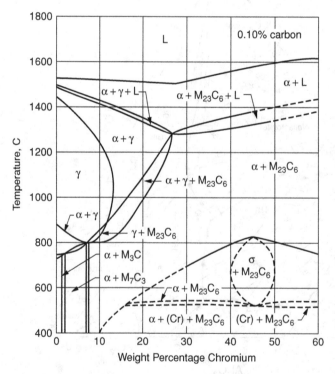

6.8. Section at 0.10% C of the iron-rich end of the Fe–Cr–C phase diagram. Reprinted with permission of ASM International® from American Society for Metals, *Metals Handbook*, 8th ed., vol. 8, *Metallography, Structures, and Phase Diagrams* (Materials Park, OH: ASM, 1973). All rights reserved. www.asminternational.org.

PHASE DIAGRAMS

lines between compositions in equilibrium with one another do not in general lie in the plane of the cut. It is, therefore, impossible to determine from a vertical section the compositions or the amounts of the phases in equilibrium.

Figure 6.8 is the section of the Fe–Cr–C diagram at 0.10% C. Note that there is an $\alpha + \gamma$ two-phase region but no α single-phase region. The α phase cannot dissolve at 0.10% C.

NOTES OF INTEREST

1. Josiah Gibbs (1839–1903) was the first American scientist after Benjamin Franklin to establish an international reputation. He earned the first engineering PhD granted in the United States from Yale in 1863. His dissertation was titled *On the Form of Teeth in Wheels in Spur Gearing*. After studying at various places in Europe, he returned to teach and do research at Yale (without salary). He published his work on thermodynamics and statistical mechanics in the *Transactions of the Connecticut Academy of Arts and Sciences*. This journal was little read in the United States, but he sent reprints to the leading scientists in Europe, where his reputation grew rapidly. In 1876, Clerk Maxwell said of Gibbs, "an obscure American has shown that the problem which long has resisted the efforts of myself and others may be solved at once."

2. The equilibrium crystallographic form of ice depends on temperature and pressure. A phase diagram showing some of these is shown in Figure 6.9.

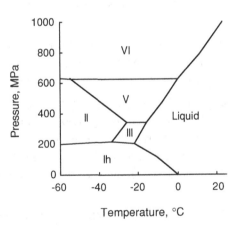

6.9. Phase diagram for ice. Ice Ih is hexagonal, ice II is rhombohedral, ice III and ice VI are tetragonal, and ice V is monolinic. There are five other forms of ice.

REFERENCES

A. Finlay, A. N. Campbell, and N. O. Smith. *The Phase Rule and Its Applications*. New York: Dover, 1951.

P. Haasen. *Physical Metallurgy*, 2nd ed. Cambridge, U.K.: Cambridge Univ. Press, 1986.

D. A. Porter and K. E. Easterling. *Phase Transformations in Metals and Alloys*, 2nd ed. London: Chapman and Hall, 1992.

PROBLEMS

1. A portion of the isothermal section of an aluminum–iron–manganese phase diagram at 600 °C is shown in Figure 6.10. Assuming equilibrium, list the phases present, give their compositions, and calculate the relative amounts of them for
   A. 0.40% Fe, 0.40% Mn and
   B. 0.20% Fe, 0.60% Mn.

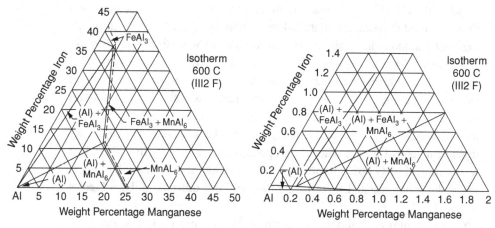

6.10. The Al–Fe–Mn phase diagram. The aluminum-rich corner is enlarged at the right. Reprinted with permission of ASM International® from American Society for Metals, *Metals Handbook*, 8th ed., vol. 8, *Metallography, Structures, and Phase Diagram* (Materials Park, OH: ASM, 1973). All rights reserved. www.asminternational.org.

2. Four distinct phases were observed in the microstructure of a ternary alloy at room temperature. Discuss this observation briefly in terms of the phase rule. What is the most likely explanation?

3. Consider the freezing of a ternary eutectic. The pressure is constant. The liquid simultaneously freezes to three solid phases, so there are four phases present during the freezing. One student applies the phase rule and concludes that there are zero degrees of freedom. Another student says that this is wrong because the amounts of the phases are not constant. Who is right? Discuss briefly.

4. Considering both temperature and pressure to be variables, what is the largest number of phases that can coexist at equilibrium in a binary alloy?

5. Consider a binary alloy of copper and silver. At a temperature near 850 °C and an alloy composition near 80% Cu–20% Ag, the phase diagram says that there should be a copper-rich solid and a liquid. Considering the pressure to be fixed at one atmosphere but allowing temperature variations,

how many degrees of freedom are there? What is (are) the independent variable(s)?

6. Trace the liquid composition during freezing of 50% SiO_2, 25% Al_2O_3, and 25% CaO. See Figure 6.6. What is the composition of the last liquid to solidify? What phases are present in the final solid?

7 Free Energy Basis of Phase Diagrams

Gibbs free energy

Phase diagrams are related to how the Gibbs free energy of a system varies with composition, temperature, and pressure. Equilibrium corresponds to the state of lowest free energy. The Gibbs free energy of a system, G, is defined as $G = H - TS$, where H is the enthalpy of the system, S is the entropy of the system, and T is the temperature. The enthalpy or heat content is given by

$$H = \int C dT + \Sigma(\Delta H_{\text{trans}}), \tag{7.1}$$

where the ΔH_{trans} terms are the latent heats associated with phase changes. The value of H is relative to a base temperature, that is, the lower limit of the integral.

Entropy is a measure of the randomness of a system. It is given by

$$S = \int (C/T) dT + \Sigma(\Delta H_{\text{trans}}/T). \tag{7.2}$$

In every system, equilibrium corresponds to the state of lowest free energy. When two components form a solution, the change of free energy on mixing is

$$\Delta G_m = \Delta H_m - T \Delta S_m, \tag{7.3}$$

where the subscript m refers to the change on mixing.

Enthalpy of mixing

When a solution is formed, the enthalpy may increase, decrease, or remain unchanged. If the enthalpy does not change ($\Delta H_m = 0$), the solution is called an *ideal solution*. If ΔH_m is positive, cooling occurs on mixing because heat is absorbed to form the solution. This indicates that AA and BB bonds are stronger than AB bonds. On the other hand, if there is a strong attraction between unlike near neighbors, ΔH_m is negative and heat will be released, warming the solution.

FREE ENERGY BASIS OF PHASE DIAGRAMS

For every two AB bonds formed by mixing, an AA and a BB bond must be broken. Therefore, the heat of mixing is given by

$$\Delta H_m = P_{AB}[\varepsilon_{AB} - (\varepsilon_{AA} + \varepsilon_{BB})/2], \tag{7.4}$$

where P_{AB} is the number of AB bonds per mole and ε_{AB}, ε_{AA}, and ε_{BB} are the energies of the AB, AA, and BB bonds. An estimate of P_{AB} may be made by assuming a random solution. In that case

$$P_{AB} = (n_o z/2) X_A X_B, \tag{7.5}$$

where n_o is Avogadro's number, z is the coordination number, and X_A and X_B are the mole fractions of A and B. Substituting Equation (7.5) into Equation (7.4),

$$\Delta H_m = X_A X_B \Omega, \tag{7.6}$$

where $\Omega = (n_o z/2)[\varepsilon_{AB} - (\varepsilon_{AA} + \varepsilon_{BB})/2]$.

Entropy of mixing

According to statistical mechanics, the change of entropy, ΔS_m, for a random mixture is given by

$$\Delta S_m = k \ln(p), \tag{7.7}$$

where p is the number of distinguishable arrangements of atoms and k is Boltzmann's constant. Consider a mole of atoms where n_A is the number of atoms of element A, and n_B is the number of atoms of element B. The total number of atoms is given by $n_o = n_A + n_B = 6.02 \times 10^{23}$. In a mole of solution, there are n_o possible sites for atoms in a crystal. Filling these sites one at a time, with n_a atoms of A, the first atom of A can be put in any of n_o sites, the second in any of $(n_o - 1)$ sites, the third in any of $(n_o - 2)$ sites, and so on, until the last atom of A can be put in any of $(n_o - n_a + 1)$ sites. The total ways of filling the sites is then equal to n_o, $(n_o - 1)$, $(n_o - 2) \ldots (n_o - n_a + 1)$. However, this is not the number of distinguishable arrangements because it does not matter whether we put the 3rd A atom in the 3rd site or the 1st site or the 112th site. The number of distinguishable arrangements is

$$p = [n_o \cdot (n_o - 1) \cdot (n_o - 2) \cdot \ldots (n_o - n_A + 1)]/n_A!. \tag{7.8}$$

If both the numerator and denominator are multiplied by $n_b!$, Equation (7.8) can be expressed as

$$p = n_o!/(n_A! n_B!) \tag{7.9}$$

so

$$\Delta S_m = k \ln(p) = k[\ln(n_o!) - \ln(n_A!) - \ln(n_B!)]. \tag{7.10}$$

Using Stirling's approximation, $\ln(x!) = x \ln x - x$,

$$\Delta S_m = k[n_o \ln(n_o) - n_a \ln(n_a) - n_b \ln(n_b) - n_o + n_A + n_B]. \tag{7.11}$$

Simplifying, $\Delta S_m = k(n_A + n_B) \ln(n_o) - n_A \ln(n_A) - n_B \ln(n_B)]$, or $\Delta S_m = -k[n_o \ln(n_A/n_o) + n_o \ln(n_B/n_o)]$. Now, recognizing that the mole fraction of element A is $N_A = n_A/n_o$, the mole fraction of element B is $(1 - N_A) = N_B = n_B/n_o$, and the gas constant $R = kn_o$,

$$\Delta S_m = -R[N_A \ln N_A + N_B \ln N_B]. \tag{7.12}$$

Equation (7.12) is based on a random solution. If the solution is not random, ΔS_m will be somewhat lower. However, Equation (7.10) is a good approximation.

The change of Gibbs free energy on mixing is then

$$\Delta G_m = \Delta H_m - T \Delta S_m = \Delta H_m + RT[X_A \ln X_A + X_B \ln X_B]. \tag{7.13}$$

Note that the second term is always negative since both X_A and X_B are both less than one and the natural log of a number less than one is negative. ΔH may be either negative or positive so ΔG can also be either negative or positive. Substituting $\Delta H_m = X_A X_B \Omega$,

$$\Delta G_m = X_A X_B \Omega + RT[X_A \ln X_A + X_B \ln X_B]. \tag{7.14}$$

Elements A and B are completely miscible if Ω is negative. However, the solubility is limited if Ω is positive. Figure 7.1 is a plot of Equation (7.14) for a positive value of Ω. The solubilities correspond to the tangent points for a line tangent to both minima.

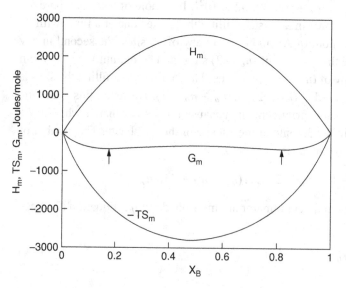

7.1. The variation of H_m, TS_m, and G_m with composition for a system with limited solubility. The arrows indicate the solubility limits of the terminal solid solutions.

Solid solubility

The solubility limit of B corresponds to $d(\Delta G_m)/dX_B = 0$. Expressing Equation (7.14) as $\Delta G_m = X_B(1 - X_B)\Omega + RT[(1 - X_B)\ln(1 - X_B) + X_B \ln X_B]$ and differentiating, $d(\Delta G_m)/dX_B = \Omega(1 - 2X_B) + RT[-(1 - X_B)\ln(1 - X_B) - \ln(1 - X_B) + X_B \ln X_B + \ln X_B] = 0$. Therefore, the solubility of B corresponds to $\Delta H_m = X_A X_B \Omega$:

$$RT[(1 + X_B)\ln X_B - \ln(1 - X_B)(2 - X_B)] = \Omega(1 - 2X_B). \quad (7.15)$$

For dilute solutions $X_B \to 0$, so $\ln(1 - X_B) \to 0$, $1 + X_B \to 1$, and $(1 - 2X_B) \to 1$, so $RT \ln X_B \to -\Omega$. The solubility is given by

$$X_B \approx \exp[-\Omega/(RT)]. \quad (7.16)$$

Equation (7.16) is a good approximation to the solvus in many systems. It can also be used to predict the equilibrium solubilities of vacancies and interstitial defects.

EXAMPLE 7.1: Find the solubility of B in A at 860 °C if $\Omega = -30000$ J/mole.

SOLUTION: Substituting $T = 860\,°C = 1133\,K$ and $R = 8.314$ into Equation (7.14),

$$9216RT[(1 + X_B)\ln X_B - \ln(1 - X_B)(2 - X_B)]/(1 - 2x_B) = \Omega.$$

Solving by trial and error, $X_B = 0.079$.

Relation of phase diagrams to free energy curves

If total free energy of each phase is plotted against composition, the free energy of the system in the two-phase region is given by the line tangent to the free energies of the two phases. Furthermore, the solubility in each phase is given by its tangent point. Figure 7.2 shows the relation between a free energy versus composition plot and the phase diagram for a system that forms a single solid solution.

Figure 7.3 shows the free energy versus composition curves for a system that forms a simple eutectic. At the lowest temperature, T_1, there is equilibrium between two solid solutions, α and β. At the eutectic temperature, T_2, there is a common tangent between α, liquid, and β. At T_3, equilibrium corresponds to G_α, the tangent between G_α and G_L, G_L, the tangent between G_L and G_β, and G_β. Finally, at T_4, the lowest free energy corresponds to liquid for all compositions.

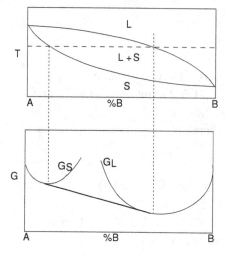

7.2. Relation of free energy diagrams to a phase diagram at a specific temperature.

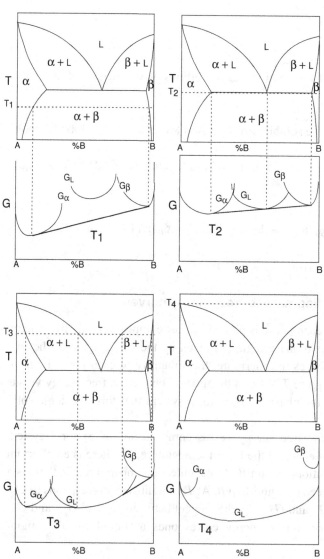

7.3. Relation of the free energy diagrams to the phase diagram for a eutectic system at four different temperatures.

FREE ENERGY BASIS OF PHASE DIAGRAMS

Pressure effects

Le Chatelier's principle states that a system will shift in the direction that nullifies the effect of a pressure change. Increased pressure favors the denser phase. Figure 7.4 shows the effects of temperature and pressure on water. Note that because liquid water is denser than ice, increased pressure lowers the melting point. It is said that forces required for ice skating are very low because the pressure of the skates on the ice creates a liquid film. At constant temperature,

$$(\partial G_{L \to S}/\partial P)_T = \Delta V_{L \to S}, \tag{7.17}$$

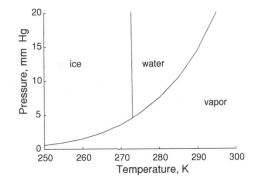

7.4. The phase diagram for water.

where G and V are the free energy and volume per mole. Consequently,

$$dT_{eq}/dP = T_{eq}\Delta V/\Delta H, \tag{7.18}$$

where T_{eq} is the equilibrium phase transformation temperature, ΔV is the transformation volume change per mole, and ΔH is the latent heat of transformation. This is known as the Clausius–Clapeyron equation. For example, for water $T_{eq} = 273$ K, $\Delta V = 1.8 \times 10^{-6}$ m^3/mol, $\Delta H = 6$ kJ/mol, so a pressure increase of one atmosphere (10 kPa) will decrease the melting temperature by $(10,000)(273)(1.8 \times 10^{-6}$ m^3/mol$)/6 \times 10^3$ J/m$^3 = 8.28 \times 10^{-4}$ °C.

Metastability

Sometimes a phase will appear when the presence of another phase would lower the free energy. The presence of cementite (Fe$_3$C) in iron–carbon alloys is an example. True equilibrium in iron–carbon alloys involves graphite rather than cementite. Figure 7.5 shows the iron–carbon diagram. The dotted lines represent the true equilibrium between austenite and graphite and between ferrite and graphite. The solid lines show the metastable equilibrium with cementite. The free energy curves in Figure 7.6 illustrate why the solubility of carbon in austenite and ferrite is lower for the true equilibrium with graphite than it is for the metastable equilibrium with cementite.

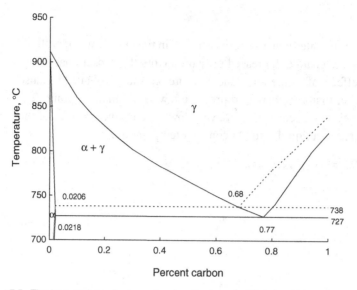

7.5. The iron–carbon diagram showing the true equilibrium with graphite (dotted lines) as well as the metastable equilibrium with cementite.

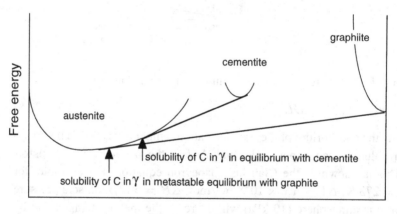

7.6. The solubility of carbon in austenite. If there is true equilibrium with graphite, the solubility is lower because the point of tangency on the austenite free energy curve is lower than for metastable equilibrium with cementite.

When austenite is transformed to ferrite and pearlite below 727 °C, the composition of the pearlite and the amount of proeutectoid ferrite depend on the transformation temperature. The reason for this can be understood by extrapolating below 727 °C the line that represents the solubility of carbon in austenite, as shown in Figure 7.7. In a steel that contains less than 0.77% C, proeutectoid ferrite must form before any pearlite forms. Ferrite formation enriches the carbon content of the austenite. Pearlite can form only when the austenite has been enriched enough so that it is saturated with respect to carbon. This happens at 0.77% C if the transformation occurs at 727 °C. At temperatures below 727 °C,

FREE ENERGY BASIS OF PHASE DIAGRAMS

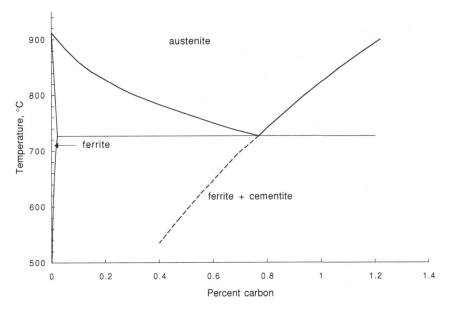

7.7. Extrapolation of the line that represents the solubility of carbon in austenite.

the carbon solubility in metastable austenite is lower. At 600 °C, cementite will first form until the composition of the austenite reaches about 0.5% C, so the pearlite will contain 0.5% C instead of 0.77%.

EXAMPLE 7.2: Determine the ratio of the widths of ferrite and cementite in pearlite that has been formed at 727 and 600 °C. Assume that the densities of cementite and ferrite are equal and the composition of cementite is 6.67% C.

SOLUTION: Pearlite formed at 727 °C will contain 0.77% C. Using the lever law on the pearlite composition,

$$f_\alpha = (6.67 - 0.77)/(6.67 - 0.02) = 0.887,$$
$$f_{cem} = 0.113, \quad \text{so} \quad w_\alpha/w_{cem} = 7.9.$$

Pearlite formed at 600 °C will contain about 0.50% C. Using the lever law on the pearlite composition, $f_\alpha = (6.67 - 0.50)/(6.67 - 0.0) = 0.925$, $f_{cem} = 0.075$, so $w_\alpha/w_{cem} = 12.3$.

The effect of this extrapolation can be seen in isothermal diagrams. Figure 7.8 is the isothermal transformation diagram for a 1050 steel. Note that proeutectoid ferrite must form before cementite for transformation temperatures above about 600 °C in accordance with Figure 7.8. The agreement is not perfect because in addition to 0.50% C, the 1050 steel contains 0.91% Mn, which lowers the eutectoid temperature and composition.

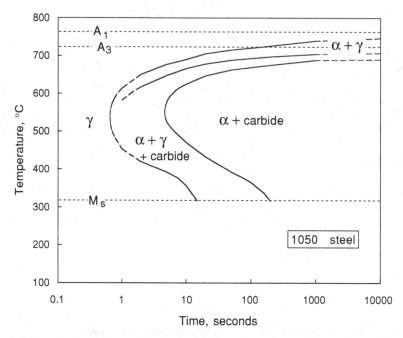

7.8. Isothermal transformation diagram for a 1050 steel.

Extrapolations of solubility limits

Extrapolation of a phase boundary between α and β that represents the solubility B into the two-phase region $\alpha + \gamma$ represents the solubility of B in α that is in metastable equilibrium (see Figure 7.9A). The solubility of the metastable phase must be greater than for the true equilibrium. If a phase diagram were drawn so that the extrapolation of the phase boundary between α and β extended into the α single-phase region (Figure 7.9B), this would predict that the solubility of the metastable phase was less than for the true equilibrium. This is clearly

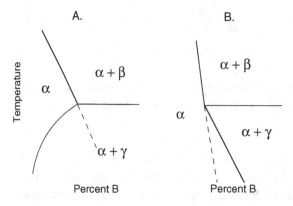

7.9. The extrapolation of the boundary of a single-phase region must not extend into that single-phase region. Diagram A is possible. Diagram B is not.

FREE ENERGY BASIS OF PHASE DIAGRAMS

impossible. Generalizing, the phase boundaries must meet at angles less than 180°.

NOTES OF INTEREST

1. Henri Louis Le Chatelier's (1850–1936) first education in chemistry and mathematics came from his engineer father who was involved in starting France's aluminum industry. Later he graduated from the École Polytechnique with the intention of becoming a mining engineer. His study of phase equilibrium led to his first proposal in 1884 of the principle which bears his name.* Later he rephrased it as "Every change of one of the factors of an equilibrium occasions a rearrangement of the system in such a direction that the factor in question experiences a change in a sense opposite to the original change." He was no doubt heavily influenced by Willard Gibbs, whose works he first translated into French.

2. It is commonly believed that ice is slippery because the pressure from our weight causes the ice under our feet to melt. However, even the concentrated pressure under ice skates lowers the melting temperature only a few degrees Celsius. One would not be able to skate when the temperature was 25 °F (−4 °C). The real explanation is that even without pressure there is a thin film of water, a few molecules thick, on the surface of ice, as illustrated in Figure 7.10. The reason is that the molecules on the surface are bound to fewer neighboring molecules than those in the interior, and therefore vibrate more. This surface film persists for several tens of degrees below the equilibrium melting point. Similar liquid films have been observed on lead just below its melting point. This film had been postulated by Faraday and Tyndall as early as 1842, but James and William Thompson (Lord Kelvin) countered with the argument that the thin water layer was caused by pressure. Only recently has there been proof of the liquid film without pressure.

7.10. Liquid layer a few molecules thick on the surface of ice. From J. G. Dash and J. W. Wettlaufer, *Scientific American* 282 (2005): 50–3.

---

* H. L. Le Chatelier, *Comptes rendus* 99 (1884).

REFERENCES

L. S. Darken and R. W. Gurry. *Physical Chemistry of Metals*. New York: McGraw-Hill, 1953.

D. Gaskell. *Introduction to Metallurgical Thermodynamics*, 3rd ed. New York: Taylor & Francis, 1995.

P. Haasen. *Physical Metallurgy*, 2nd ed. Cambridge, U.K.: Cambridge Univ. Press, 1986.

W. F. Hosford. *Physical Metallurgy*. Boca Raton, FL: CRC Press, 2005.

D. A. Porter and K. E. Easterling. *Phase Transformations in Metals and Alloys*, 2nd ed. London: Chapman & Hall, 1992.

J. S. Wettlaufer and J. G. Dash. *Ice Physics and the Natural Environment*. New York: Springer-Verlag, 1999.

PROBLEMS

1. Figure 7.11 shows the free energy versus composition curves for the AB system. Construct the AB phase diagram.

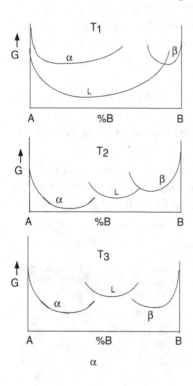

7.11. Free energy versus composition diagrams for the AB system at three temperatures, $T_1 > T_2 > T_3$.

2. In the CD system, the C-rich solid solution, γ, is often in metastable equilibrium with C_2D, but true equilibrium exists between γ and the compound, CD_2. Compare the corresponding solubilities of D in γ.

3. Sketch a plausible free energy versus composition diagram for the AB system in Figure 7.12 at temperature T.

FREE ENERGY BASIS OF PHASE DIAGRAMS

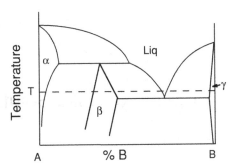

7.12. The AB phase diagram.

4. Estimate the change of the melting point of aluminum caused by a pressure change of 1 kbar. The density of solid aluminum is 2.7 Mg/m^3, its melting point is 660 °C, its heat of fusion is 98 kJ/mol, and aluminum contracts 6% when it freezes.

5. The solubility of carbon in iron is 0.0218 at 727 °C. Estimate the solubility at 200 °C.

6. Use the Clausius–Clapeyron equation to calculate the pressure required to depress the melting point of ice by 1 °C. The density of ice is 917 kg/m^3, the density of water is 1000 kg/m^3, and the heat of fusion is 3.3×10^5 J/kg.

7. Estimate the composition of pearlite formed in an Fe–C alloy isothermally at 700 °C. What is the ratio of Fe$_3$C to α in this pearlite?

8. Estimate the lowering of the melting point of ice by the weight of a 180-lb skater. Assume that his entire weight is on one skate and that the skate makes an area of contact of 100 mm^2 with the ice.

8 Ordering of Solid Solutions

Solid solutions are not usually random. If ΔH_m is positive, there is a tendency to form fewer AB bonds than would exist in a random solution. The result is *clustering*. On the other hand, if ΔH_m is negative, there is a tendency to form more AB bonds than would exist in a random solution and the result is *ordering*. Ordering may be either long range or short range.

Long-range order

Long-range order is possible in some binary solid solutions having compositions corresponding to simple ratios of the number of A and B atoms. One species of atom may tend to occupy certain lattice positions. Figure 8.1 shows several ordered structures and Table 8.1 lists compositions that can form these ordered structures.

Alloys of copper and platinum form two other ordered structures. The composition CuPt forms an ordered structure with copper and platinum atoms occupying alternating close-packed planes in what would otherwise be an fcc solid solution. The ordering in the composition Cu_3Pt_5 can be visualized as alternating close-packed planes, one filled with Pt atoms and the other having Pt atoms in one fourth of the sites.

There is a long-range order parameter, s, defined as the fraction of A atoms, f_A, occupying the correct sites minus the fraction occupying the wrong sites:

$$s = f_A - (1 - f_A) = 2f_A - 1. \tag{8.1}$$

This parameter varies from 0 for a random solid solution to 1 for perfect order. The degree of order decreases with increasing temperature and drops to zero at a critical temperature, T_c. The temperature dependence of s is shown in Figure 8.2.

Long-range order domains may be nucleated at several places within a grain. When the domains grow together, a boundary will be formed if the domains are out of phase. Figure 8.3 illustrated such an antiphase domain boundary. The

ORDERING OF SOLID SOLUTIONS

Table 8.1. Ordered structures

L2$_0$	L1$_2$	L1$_0$	DO$_3$	DO$_{19}$
CuZn	Cu$_3$Au	CuAu	Fe$_3$Al	Mg$_3$Cd
FeCo	Au$_3$Cu	CoPt	Fe$_3$Si	Cd$_3$Mg
NiAl	Ni$_3$Mn	FePt	Fe$_3$Be	Ti$_3$Al
FeAl	Ni$_3$Al		Cu$_3$Al	Ni$_3$Sn
AgMg	Pt$_3$Fe			

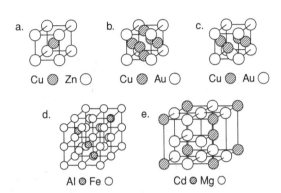

8.1. Five common ordered structures: (A) L2$_0$-type CuZn, (B) L1$_2$-type Cu$_3$Au, (C) L1$_0$-type CuAu, (D) DO$_3$-type Fe$_3$Al, and (E) DO$_{19}$-type Mg$_3$Cd. From W. F. Hosford, *Physical Metallurgy* (Boca Raton, FL: CRC Press, 2004), p. 97, figure 5.6.

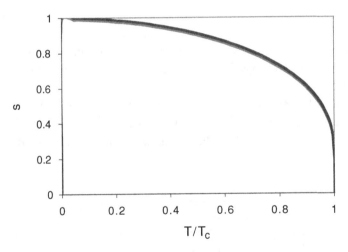

8.2. Long-range order parameter, s, decreases with temperature.

L2$_0$ ordering corresponds to the body-centered cubic with either of the species occupying the body-centered sites. In this case there are two possible "phases" to the ordering. In the L1$_2$ ordering the Au atoms may occupy any of four equivalent sites, so there are four "phases."

The change from an ordered state to a disordered state is a second-order phase change. In a first-order phase change, thermodynamic properties such as enthalpy and entropy undergo abrupt changes. In contrast, there is no heat of

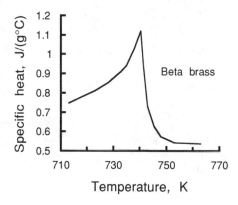

8.3. The antiphase boundary between two domains.

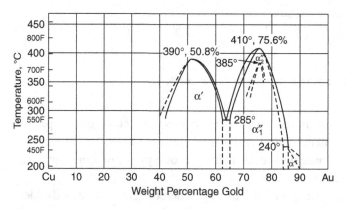

8.4. Excess specific heat of β brass near the critical temperature.

transformation, ΔH, but there is an excess specific heat, as illustrated in Figure 8.4 for β brass. This is similar to the change from ferromagnetic to nonferromagnetic states in iron at the Curie temperature.

Phase diagrams show the compositions for which ordering is possible. With increased temperature, the range of compositions decreases to the stochiometric composition at the Curie temperature. The phase diagram for the CuAu system is given in Figure 8.5.

8.5. Low-temperature region of the Cu–Au phase diagram showing the ordered phase regions, Cu_3Au (α') and CuAu (α''). Reprinted with permission of ASM International® from *ASM Handbook*, vol. 8, 8th ed. (Materials Park, OH: ASM, 1973), p. 267. All rights reserved. www.asminternational.org.

ORDERING OF SOLID SOLUTIONS

Effect of long-range order on properties

The effect of long-range order on the electrical resistivity of copper–gold alloys is shown in Figure 8.6. The increased periodicity produced by ordering during annealing at 200 °C decreases the resistivity.

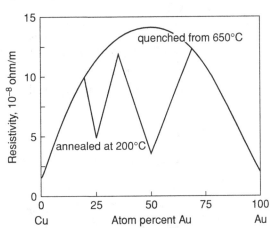

8.6. Ordering lowers the electrical resistivity of copper–gold alloys. Adapted from C. S. Barrett, *Structure of Metals and Alloys* (New York: McGraw-Hill, 1943), p. 244, figure 14.

Long-range order also increases the yield strength. Passage of a normal dislocation through a structure with long-range order produces an antiphase domain boundary. The introduction of an antiphase domain boundary increases the system's energy so a greater force is required than to move a dislocation through a disordered structure.

Short-range order

Even in the absence of long-range order, the positions of atoms in substitutional solid solutions may not be random. If the strength of AB bonds is greater than average of the AA and BB bond strength, A atoms will tend to be surrounded by B atoms. This is called *short-range order*. If, on the other hand, the average of the AA and BB bond strengths is greater than that of an AB bond, there will be clustering. An A atom will have more than the statistical number of B atoms. The degree of short-range order can be characterized by a parameter,

$$\sigma = (N_{AB} - N_{AB\mathrm{random}})/(N_{AB\mathrm{max}} - N_{AB\mathrm{random}}). \tag{8.2}$$

For a random solid solution, $\sigma = 0$. Perfect order corresponds to $\sigma = 1$. A negative value of σ indicates clustering. The tail of excess specific heat above the Curie temperature in Figure 8.4 reflects the loss of remaining short-range order.

NOTE OF INTEREST

In 1919, G. Tammann[*] predicted that long-range order might exist in alloys. He used the German term *Überstructur*, which translates as "superlattices," to describe the phenomenon. Papers by C. H. Johannson, J. O. Linde, and G. Borelius

---

[*] G. Tammann, *Zeit. anorg. Chem.* 1 (1919).

starting in 1925 reported on ordering in gold–copper, copper–palladium, and copper–platinum. These provided much of the early understanding of long-range order.

REFERENCES

C. S. Barrett. *Structure of Metals and Alloys*. New York: McGraw-Hill, 1943.
H. Cottrell. *Theoretical Structural Metallurgy*. London: Edward Arnold, 1948.
D. A. Porter and K. E. Easterling. *Phase Transformations in Metals and Alloys*, 2nd ed. London: Chapman & Hall, 1992.

PROBLEMS

1. Calculate the value of s for the two-dimensional solutions shown in Figure 8.7A and B.

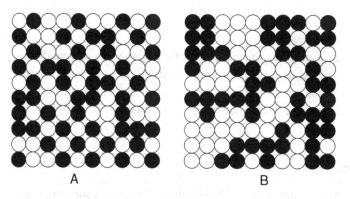

8.7. An AB solid solution with $N_A = N_B = 50$.

2. How many phases are there for $L1_2$-type Cu_3Au ordering?
3. Sketch a simple two-dimensional crystal that has been ordered. Now make another sketch, showing the change that occurs by passage of a single dislocation.
4. Discuss the possibility of order in interstitial solid solutions.
5. Sketch the close-packed planes in Cu_3Pt_5.

9 Diffusion

Fick's first law

Fick's first law states that in a solution with a concentration gradient there will be a net flux of solute atoms from regions of high solute concentration to regions of low concentration and that the net flux of solute is proportional to the concentration gradient. This can be expressed as

$$J = -D\,dc/dx, \tag{9.1}$$

where J is the net flux of solute, D is the diffusivity (or diffusion coefficient), c is the concentration of solute, and x is distance. See Figure 9.1.

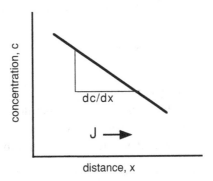

9.1. The diffusion flux, J, is proportional to the concentration gradient. Note that the gradient shown is negative ($dc/dx < 0$) and the flux, J, is positive.

The flux is the net amount of solute crossing an imaginary plane per area of the plane and per time. The flux may be expressed as solute atoms/(m² · s), in which case the concentration, c, must be expressed as atoms/m³. Alternatively, J and c may be expressed in terms of mass of solute, the units of J being (kg solute)/(m² · s) and of c being (kg solute)/m³. The diffusivity has dimensions of m²/s and depends on the solvent, the solute, the concentration, and the temperature.

Fick's first law for mass transport by diffusion is analogous to the laws of thermal and electrical conduction. For heat conduction, $q = k\,dT/dx$, where dT/dx is a thermal gradient (°C/m); k is the thermal conductivity, J/(ms°C); and q is the flux;

J/(m²s). Ohm's law, $I = E/R$, can be expressed in terms of a current density, $i = \sigma \varepsilon$, where the current density, i, in coulombs/(m²s) is a flux; ε is the voltage gradient (V/m); and the proportionality constant is the conductivity, $\sigma = 1/\rho$, in (ohm · m)$^{-1}$.

Direct use of Fick's first law is limited to steady-state (or nearly steady-state) problems in which the variation of dc/dx over the concentration range of concern can be neglected.

Fick's second law

Fick's second law expresses how the concentration at a point changes with time. According to Fick's first law the flux into an element of unit area and thickness, dx, is $J_{in} = -Ddc/dx$ and the flux out of it is $J_{out} = -Ddc/dx - d(-Ddc/dx)$ (see Figure 9.2). The rate of change of the composition within the element is then $dc/dt = J_{in} - J_{out}$ or

$$dc/dt = \partial(Ddc/dx). \tag{9.2}$$

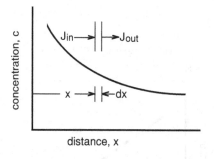

9.2. The rate of change of composition in an element of volume, Adx, equals the differences between the fluxes into and out of the element.

This is a general statement of Fick's second law, which recognizes that the diffusivity may be a function of concentration and therefore of distance, x. In applications where the variation of D with distance and time can be neglected, Equation 9.2 can be simplified into a more useful form:

$$dc/dt = Dd^2c/dx^2. \tag{9.3}$$

Rigorously, concentration should be expressed in atoms or mass per volume, but if density changes are neglected concentration may be expressed in atomic percent or weight percent.

Solutions of Fick's second law and the error function

There are specific solutions to Fick's second law for specific boundary conditions.

Addition of material to or removal from a surface: If the composition at the surface of a material is suddenly changed from its initial composition, c_o, to a

DIFFUSION

new composition, c_s, and held at that level (Figure 9.3), the solution to Equation (9.3) is

$$c = c_s - (c_s - c_o)\,\text{erf}[x/(2\sqrt{Dt})], \tag{9.4}$$

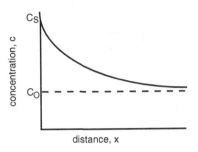

9.3. Solution to Fick's second law for a constant surface concentration, c_s.

where erf is the error function defined as

$$\text{erf}(x) = (2/\pi^{1/2}) \int_0^x \exp(-t^2)\,dt. \tag{9.5}$$

Table 9.1 and Figure 9.4 show how the erf(x) depends on x. One application of this solution involves carburizing and decarburizing of steels. Three straight lines make a rough approximation to the error function, as shown in Figure 9.5. For $x \geq 1$, erf(x) = 1; for $1 \geq x \geq -1$, erf(x) = x; for $-1 \geq x$, erf(x) = -1.

Table 9.1. Values of the error function, $\text{erf}(x) = (2/\pi^{1/2}) \int_0^x \exp(-t^2)\,dt$

x	erf(x)	x	erf(x)	x	erf(x)	x	erf(x)
0.00	0.000	0.05	0.0564	0.10	0.1125	0.15	0.1680
0.20	0.2227	0.25	0.2763	0.30	0.3286	0.35	0.3794
0.40	0.4284	0.45	0.4755	0.50	0.5205	0.55	0.5633
0.60	0.6039	0.65	0.6420	0.70	0.6778	0.75	0.7112
0.80	0.7421	0.85	0.7707	0.90	0.7970	0.95	0.8209
1.00	0.8427	1.10	0.8802	1.20	0.9103	1.30	0.9340
1.40	0.9523	1.50	0.9661	1.60	0.9763	1.70	0.9838
1.80	0.9891	1.90	0.9928	2.00	0.9953	2.20	0.9981
2.40	0.9993	2.60	0.9998	2.80	0.9999		

Note: erf($-x$) = $-$erf(x), i.e., erf(-0.20) = -0.2227. For small values of x, erf(x) ≈ $2x/\sqrt{\pi}$.

Junction of two solid solutions: Another simple solution is for two blocks of differing initial concentrations, c_1 and c_2, that are welded together. In this case

$$c = (c_2 + c_1)/2 - [(c_2 - c_1)/2]\,\text{erf}[x/\sqrt{Dt}]. \tag{9.6}$$

Figure 9.6 illustrates this solution. Note that Equation (9.6) is similar to Equation (9.4), except that $(c_1 + c_2)/2$ replaces c_s and $(c_1 - c_2)/2$ replaces $c_s - c_o$.

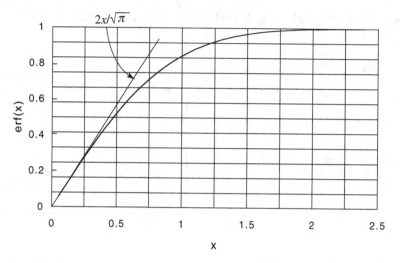

9.4. The dependence of erf(x) on x.

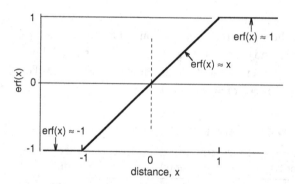

9.5. Approximation of the erf(x). For $x \leq -1$, erf(x) = -1. For $x \geq 1$, erf(x) = 1. For $-1 \leq x \leq 1$, erf(x) = x.

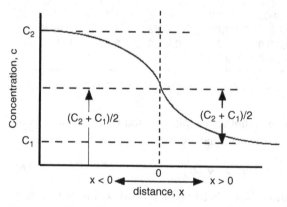

9.6. The solution of Fick's second law for two solutions with different concentrations.

DIFFUSION

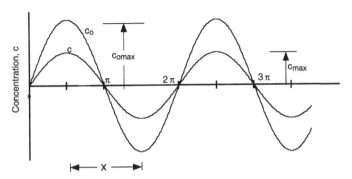

9.7. Sinusoidal concentration profile, c_o, resulting from interdendritic segregation, and the profile, c, after some homogenization.

Homogenization: Interdendritic segregation during solidification causes local composition variations that can be approximated by a sine wave of wavelength $2x$ and amplitude c_{omax}. Homogenization by diffusion decreases the amplitude c_{omax} to c_{max}, as shown in Figure 9.7. Defining c and c_o as the differences between local concentrations and the average composition, the extent of homogenization is described by

$$c/c_o = \exp[-(\pi X)^2], \qquad (9.7)$$

where $X = x\sqrt{Dt}$.

General: All solutions to Fick's second law are of the form

$$f(concentrations) = x/\sqrt{Dt}, \qquad (9.8)$$

where $f(concentrations)$ depends on the concentration at a specific point, c; the initial concentration, c_o; the surface concentration, c_s; and so on. In many problems, these concentrations are fixed so

$$x/\sqrt{Dt} = \text{constant}. \qquad (9.9)$$

Mechanisms of diffusion

Diffusion in interstitial solid solutions occurs by interstitially dissolved atoms jumping from one interstitial site to another. For an atom to move from one interstitial site to another, it must pass through a position where its potential energy is a maximum. The difference between the potential energy in this position and that in the normal interstitial site is the activation energy for diffusion and must be provided by thermal fluctuations. The overall diffusion rate is governed by an Arrhenius-type rate equation,

$$D = D_o \exp(-E/kT), \qquad (9.10)$$

where D_o is a constant for the diffusing system, k is Boltzmann's constant, T is the absolute temperature, and E is the activation energy (the energy for a single jump). Often this equation is written as

$$D = D_o \exp(-Q/RT), \qquad (9.11)$$

Table 9.2. Diffusivities for interstitials

Solvent	Solute	D_o(m²/s)	Q(kJ/mol)
Ta	O	4×10^{-7}	106.
	N	6	158.
	C	6.7	162.
Fe(γ)	C	4	80.3
	N		75.9
Ni	H	7.8	41.2
Pt	H	6	27.

Source: Reprinted with permission of ASM International® from D. N. Besher, *Diffusion* (Materials Park, OH: ASM, 1973), pp. 218–19. All rights reserved. www.asminternational.org.

where the activation energy, $Q = n_o E$, is for a mole of jumps. (n_o is Avogadro's number: 6.02×10^{23} jumps.) Correspondingly, R ($= n_o$k) is the gas constant. Experimental data for diffusion of interstitials in several metals is given in Table 9.2.

Kirkendall effect

In early studies of diffusion it was assumed that in substitutional solid solutions both species of atoms diffuse in opposite directions at the same velocity. It was assumed that there was an interchange mechanism with two atoms changing place or a cooperative rotation of a ring of four or six or more atoms, as sketched in Figure 9.8. However, experiments by Smigelskas and Kirkendall[*] showed that this could not be so. They studied diffusion in a diffusion couple formed by plating copper onto a brass bar containing 30% zinc (Figure 9.9). Molybdenum wires were wrapped around the bar before plating so the initial interface could be located after diffusion. Examination of the couple after diffusion revealed an apparent movement of the wires. That is, the distance between the wires and center of the bar had decreased. Because the wires cannot diffuse, the only reasonable interpretation is that the net flux of zinc past the wires in one direction was faster than the flux of copper atoms in the opposite direction. This observation is inconsistent with all of the exchange mechanisms for diffusion and thereby provided strong supporting evidence for the vacancy mechanism of diffusion.

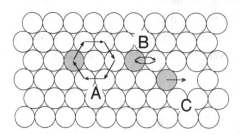

9.8. Schematic illustration of several mechanisms proposed for substitutional diffusion. (A) Ring interchange, (B) simple interchange, and (C) vacancy migration.

[*] A. Smigelskas and E. Kirkendall, *Trans. AIME* 171 (1947): 130–42.

DIFFUSION

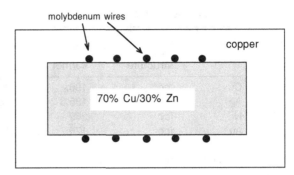

9.9. Diffusion couple in the Kirkendall experiments. Because zinc atoms diffused faster than copper atoms, the molybdenum wires appeared to move toward the center of the specimen.

Porosity can be formed because of the different rates of diffusion of two species. Because zinc diffuses faster than copper, there is a net flux of vacancies into the zinc-rich brass. Under some circumstances, these vacancies may diffuse to grain boundaries. In this case there is a volume contraction as the vacancies disappear into the boundaries. However, volume contraction of the brass may be prevented by macroscopic constraints. Then porosity will result from the precipitation of the vacancies to form voids. Such porosity can occur as concentration gradients formed by interdendritic segregation are minimized during annealing. In brass ingots, the centers of the dendrite arms are copper rich and the interdendritic regions are zinc rich. When the mechanical working of an ingot reduces the dendritic spacing enough that homogenization takes place during annealing, porosity will be formed in the zinc-rich regions.

Temperature dependence

The equilibrium number of vacancies depends exponentially on temperature:

$$n_v = n_o \exp(-E_f/kT), \tag{9.12}$$

where n_v/n_o is the fraction of the lattice sites that are vacant and E_f is the energy to form a vacancy. The rate that a given vacant site will be filled by a substitutional atom moving into it is also dependent on thermal activation,

$$\text{rate} = \exp(-E_m/kT), \tag{9.13}$$

where E_m is the energy barrier to fill a vacancy by movement of an adjacent substitutional atom. The net rate of diffusion is proportional to the product of the number of vacancies and the rate at which they contribute to diffusion. Therefore, $D = D_o \exp(-E_f/kT) \cdot \exp(-E_m/kT)$, which simplifies to

$$D = D_o \exp(-E/kT), \tag{9.14}$$

where

$$E = E_f + E_m. \tag{9.15}$$

Table 9.3. Self-diffusion data

Metal	Crystal structure	Q kJ/mol	D_o (m²/s)	T_m (K)	Q/RT_m
Cu	fcc	196.	2.32×10^{-4}	1356	17.5
Ag	fcc	185.6	1.06	1234	18.0
Ni	fcc	289	2.22	1726	19.6
Au	fcc	176.7	0.107	1336	17.8
Pb	fcc	109	1.37	600	20.4
α – Fe	bcc	240	2.01	1809	15.9
γ – Fe	fcc	267.5	0.22	1809	17.7
Nb	bcc	401	1.1	2741	19.4
Mo	bcc	460	1.8	2883	19.2
Mg	hcp	134.6	1.0	923	17.8

Source: Selected data from J. Askill, *Tracer Diffusion Data for Metals, Alloys and Simple Oxides* (New York: Plenum, 1970), pp. 31–41.

Of course, this equation can also be expressed in an equivalent form in terms of Q, the activation energy per mole of diffusion jumps:

$$D = D_o \exp(-Q/RT). \tag{9.16}$$

With the use of radioactive isotopes it has been possible to measure the rates of self-diffusion in solids. The activation energies for self-diffusion and diffusion of substitutional solutes are considerably higher than those for interstitial diffusion and therefore the diffusion rates are much lower. Data for self-diffusion in several metals are given in Table 9.3. In comparing these data, several other trends are apparent. One is that the activation energies increase with melting point. In fact, for most relatively close-packed metals (fcc, bcc, hcp), Q/T_m is nearly the same. Self-diffusion can play a significant role in such diverse phenomena as sintering, creep, thermal etching, and grain boundary migration.

Special diffusion paths

Diffusion occurs rapidly along grain boundaries, dislocations, and free surfaces. The most important of these diffusion paths are grain boundaries. Data for self-diffusion in silver are given in Table 9.4 These data are based on experiments on both polycrystals and single crystals of silver.

The overall diffusivities measured in silver are shown in Figure 9.10. It should be noted that the effects of grain boundary diffusion are observable only at low

Table 9.4. Self-diffusion in silver

Path	D_o (m²/s)	Q (kJ/mol)
Lattice	90×10^{-6}	193.0
Grain boundary*	2.3×10^{-9}	110.9

* To express the diffusion in terms of a diffusivity, an effective width of the grain boundary path was assumed to be 30 nm.

DIFFUSION

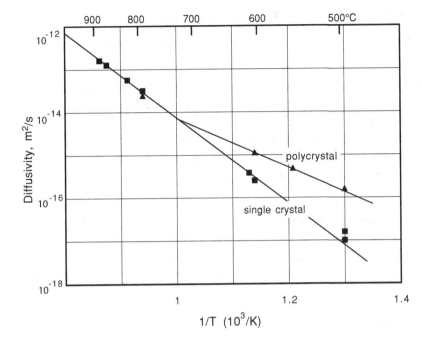

9.10. Experimental measurements of the self-diffusion coefficient in silver. Silver single crystals and polycrystals of a 35-μm grain size. The higher values of D for the polycrystal at low temperatures (high values of $1/T$) are a result of grain boundary diffusion. At high temperatures (low $1/T$), lattice diffusion masks the contribution of grain boundary diffusion.

temperatures. In most practical diffusion problems, grain boundary diffusion can be neglected.

Darken's equation

The solutions to Fick's second law (Equations (9.5)–(9.7)) are based on a single diffusivity, D, whereas the Kirkendall experiments show that each species has its own diffusivity. Darken[*] showed that Fick's second law should be written as

$$\partial N_A/\partial t = \partial/\partial x[\tilde{D}\partial N_A/\partial x], \quad (9.17)$$

where $\tilde{D}$ is the effective diffusivity and is related to the intrinsic diffusivities of the two species, D_A and D_B, by

$$\tilde{D} = N_B D_A + N_A D_B, \quad (9.18)$$

where N_A and N_B are the atomic fractions of the two species. For dilute solutions $N_B \to 0$, so $\tilde{D}$ approaches the diffusivity of the solute ($\tilde{D} \to D_B$).

Darken also showed that the velocity, v, of the markers in the Kirkendall experiments is given by

$$v = (D_A - D_B)\partial N_A/\partial x. \quad (9.19)$$

---

[*] L. S. Darken, *Trans. AIME* 175 (1948): 184–215.

Diffusion in systems with more than one phase

In analyzing diffusion couples involving two or more phases, there are two key points:

1. Local equilibrium is maintained at interfaces. Therefore, there are discontinuities in composition profiles at interfaces. The phase diagram gives the compositions that are in equilibrium with one another.
2. No net diffusion can occur in a two-phase microstructure because both phases are in equilibrium and there are no concentration gradients in the phases. These points will be illustrated by several examples.

EXAMPLE 9.1. Consider diffusion between two pure metals in a system that has an intermediate phase, as illustrated Figure 9.11. Interdiffusion between blocks of pure A and pure B at temperature T will result in the concentration profiles shown in Figure 9.12. Note that a band of β will develop at the interface. The compositions at the $\alpha - \beta$ and $\beta - \gamma$ interfaces are those from the equilibrium

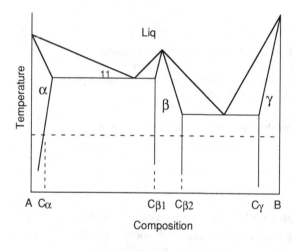

9.11. The AB phase diagram.

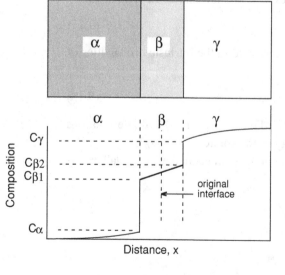

9.12. Microstructure of diffusion couple between A and B (top) and concentration profile (bottom).

DIFFUSION

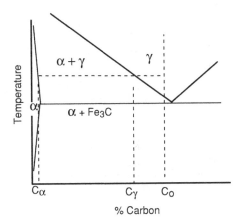

9.13. Iron–carbon diagram.

diagram so the concentration profile is discontinuous at the interfaces. No two-phase microstructure will develop.

EXAMPLE 9.2. Consider the decarburization of a steel having a carbon content of c_o when it is heated into the austenite (γ) region and held in air (Figure 9.13). At this temperature the reaction $2C + O_2 \rightarrow 2CO$ effectively reduces the carbon concentration at the surface to zero. A layer of α forms at the surface and into the steel to a depth of x. The concentration profile near the surface is shown in Figure 9.14. The concentration gradient is $dc/dx = -c_\alpha/x$, where c_α is the carbon content of the α in equilibrium with the γ. Fick's first law gives the flux, $J = -Ddc/dx = Dc_\alpha/x$. As the interface advances a distance, dx (Figure 9.15), the amount of carbon that is removed in a time interval, dt, is approximately $(c_\gamma - c_\alpha)dx$ so the flux is

$$J = (c_\gamma - c_\alpha)dx/dt. \tag{9.20}$$

Equating the two expressions,

$$(c_\gamma - c_\alpha)dx/dt = Dc_\alpha/x \tag{9.21}$$

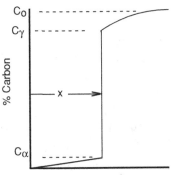

9.14. Carbon concentration profile.

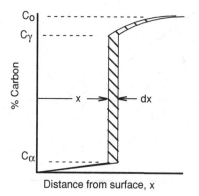

9.15. Change of concentration profile as the $\alpha - \gamma$ interface advances a distance, dx, in a time interval, dt.

so $x\,dx = D\{c_\alpha/(c_\gamma - c_\alpha)\}dt$. Integrating gives $x^2 = [2Dc_\alpha/(c_\gamma - c_\alpha)]t$ or

$$x = [2Dtc_\alpha/(c_\gamma - c_\alpha)]^{1/2}. \tag{9.22}$$

Note that for fixed concentrations $x/\sqrt{Dt}$ is constant or x is proportional to $\sqrt{Dt}$.

EXAMPLE 9.3. It is known that with a certain carburizing atmosphere it takes 8 h at 900 °C to obtain a carbon concentration of 0.75 at a depth of 0.020 in. Find the time to reach the same carbon concentration at a depth of 0.03 in. at another temperature.

SOLUTION: $x_2/\sqrt{D_2 t_2} = x_1/\sqrt{D_1 t_1}$. Let $t_1 = 8$ h, $x_1 = 0.020$ in., $x_2 = 0.030$ in., and $D_2 = D_1$. Then $t_2 = t_1(x_2/x_1)^2 = 8(.03/.02)^2 = 18$ h.

EXAMPLE 9.4. A steel containing 0.20% C is to be carburized in an atmosphere that maintains a carbon concentration of 1.20% at the surface.

A. After 10 h at 870 °C, at what depth below the surface would you find a concentration of 0.40% C? (For diffusion of C in austenite, $D_o = 2.0 \times 10^{-5}$ m²/s and $Q = 140 \times 10^3$ J/mol.)
B. How long would it take, still at 870 °C, to double the depth (part A) at which the concentration is 0.40%?
C. What carburizing time at 927 °C gives the same results as 10 h at 870 °C?

SOLUTION:

A. Using $c = c_s - (c_s - c_o)\mathrm{erf}[x/(2\sqrt{Dt})]$, with $c = 0.4$, $c_o = 0.2$, and $c_s = 1.2$, then $(c - c_o)/(c_s - c_o) = 0.2$, or $\mathrm{erf}[x/(2\sqrt{Dt})] = 0.8$. Interpolating, $x/(2\sqrt{Dt}) = 0.90 + 0.05 \cdot (0.8 - 0.7970)/(0.8209 - 0.7970) = 0.906$, so $x = 0.906 \cdot 2\sqrt{Dt}$, where $D = 2.0 \times 10^{-5} \exp[-140000/(8.31 \cdot 1143)] = 7.939 \times 10^{-12}$ m²/s, $t = 36,000$ s, and $x = 0.906 \times 2[7.939 \times 10^{-12} \times 36000]^{1/2} = 9.69 \times 10^{-4}$ m or about 1 mm.
B. $x/\sqrt{Dt} = $ constant. For the same temperature, D is fixed so $x_2/\sqrt{Dt_2} = x_1/\sqrt{Dt_1}$, or $t_2 = t_1(x_2/x_1)^2 = 10(2)^2 = 40$ h.

DIFFUSION

C. For the same carburizing results, the concentration profile must be the same. Therefore, $Dt = $ constant, or $D_2t_2 = D_1t_1$, $t_2 = t_1(D_1D_2) = 10$ h· $\exp[(-Q/R)(1/T_1 - T_2)] = 10\exp[(-140000/8.31)/(1/1143 - 1/1200)] = 4.97$ or 5 h.

EXAMPLE 9.5. A steel containing 0.25% C was heated in air for 10 h at 700 °C. Find the depth of the decarburized layer (i.e., the layer in which there is no Fe_3C). Given: The solubility of C in $\alpha -$ Fe at 700 °C is 0.016%. One may assume that the carbon concentration at the surface is negligible.

SOLUTION: At 700 °C the steel consists of two phases, α and Fe_3C. The concentration profile must appear as sketched in Figure 9.13. Near the surface there is a decarburized layer containing only α. The concentration in the α must vary from 0% C at the outside surface to $c_\alpha = 0.016$% C where it is in contact with Fe_3C. See Figure 9.16.

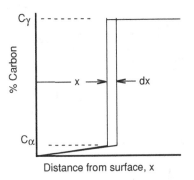

9.16. Decarburization of a steel heated in the $\alpha + Fe_3C$ phase region.

An approximate solution can be obtained by using Fick's first law to make a mass balance as the interface moves a distance of dx. The amount of carbon transported to the surface in a period, dt, is $(\bar{c}-c_\alpha)dx$ and this must equal the flux times dt, $-Jdt = D(dc/dx)dt$. Substituting, $dc/dx = (c_\alpha - 0)/x$, $(\bar{c}-c_\alpha)dx = D(c_\alpha/x)dt$ and integrating, $x^2/2 = Dtc_\alpha/(\bar{c}-c_\alpha)$, $x = [2Dtc_\alpha/(\bar{c}-c_\alpha)]^{1/2}$. Now substituting, $D = 2 \times 10^{-6}\exp[-84,400/(8.31 \times 973)] = 5.86 \times 10^{-11}$ m²/s, $\bar{c} = 0.25$, $c_\alpha = 0.016$, and $t = 36,000$ s, so $x = 0.00057$ m or 0.6 mm.

NOTE OF INTEREST

Ernest Kirkendall's doctoral thesis on the interdiffusion between copper and brass disproved the almost universally accepted concept that diffusion in substitutional solutions occurs by interchange of atoms and led to the conclusion that diffusion was the result of vacancy migration. His results were so startling to the leading metallurgists of the day that publication of his work was held up by reviewers who doubted his results.

REFERENCES

D. N. Besher. *Diffusion*. Materials Park, OH: ASM, 1973.
Diffusion in BCC Metals. Materials Park, OH: ASM, 1965.
W. F. Hosford. *Physical Metallurgy*. Boca Raton, FL: CRC Press, 2005.
P. G. Shewmon. *Diffusion in Solids*. New York: McGraw-Hill, 1963.

PROBLEMS

1. A block of an alloy of Cu – 6% Al was welded to a block of Cu – 14% Al and heated to 700 °C. Sketch the concentration profile after some diffusion occurred. Figure 9.17 shows the phase diagram.

2. Consider a piece of steel containing 0.20% C at 750 °C in an atmosphere that reduces the concentration at the surface to 0% C. $D_o = 2.0 \times 10^{-4} \text{m}^2/\text{s}$

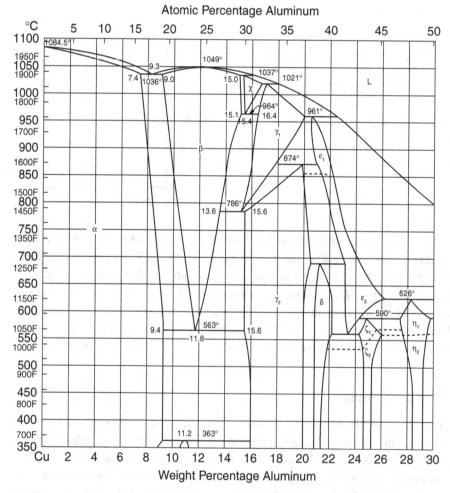

9.17. Copper–aluminum phase diagram. Reprinted with permission of ASM International® from *ASM Handbook*, vol. 8, 8th ed. (Materials Park, OH: ASM, 1973), p. 259. All rights reserved. www.asminternational.org.

and $Q = 140$ kJ/mole in $\gamma - \text{Fe}$, and $D_o = 0.2 \times 10^{-4}$ m^2/s and $Q = 84$ kJ/mol in $\alpha - \text{Fe}$.

A. What is the solubility (wt.%) of carbon in α iron at 750 °C? Sketch the concentration profile near the surface. (Plot %C vs. distance from surface.)

B. Find an appropriate diffusivity for C in iron at 750 °C.

C. Using Fick's first law, $J = -D dc/dx$, express the flux, J, in terms of the depth of the decarburized layer.

3. In an incremental time period, dt, the decarburized depth increases by dx, such that $J dt = (c_o - c_\alpha) dx$, where $J = D c_\alpha / x$.

A. Find x as a function of time by integration.

B. What would be the depth of the decarburized layer after 4 h?

For diffusion of carbon in ferrite, $Q = 84$ kJ/mol, $D_o = 0.20 \times 10^{-4}$ m^2/s.

For diffusion of carbon in austenite, $Q = 140$ kJ/mol, $D_o = 2.0 \times 10^{-4}$ m^2/s.

4. A. If the average grain diameter of silver were doubled, by what factor would the net diffusion by grain boundary diffusivity change?

B. At what temperature would the net transport by grain boundary diffusion be the same as with the original grain size at 1000 °C?

5. When iron containing 0.20% C is exposed to a carbon-bearing atmosphere at 850 °C, it is found that after 2 h, the concentration of carbon is 0.65% at 1 mm below the surface. If iron were exposed to the same atmosphere at 900 °C for 1.5 h, at what depth below the surface would the concentration of carbon be 0.65%?

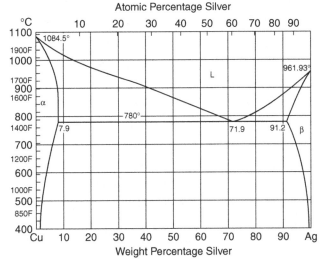

9.18. The silver–copper phase diagram. Reprinted with permission of ASM International® from L. A. Willey, *Metals Handbook*, vol. 8, 8th ed. (Materials Park, OH: ASM, 1973), p. 259. All rights reserved. www.asminternational.org.

6. Explain why one does not bother with Darken's equation when considering problems of diffusion of carbon in iron.

7. At one period, dimes and quarters were made from lamination of two alloys. The composition of the interior sheet was 90% Cu–10% Ag and the composition of the outer sheet was 10% Cu–90% Ag. Plot how the composition would vary with distance from the center after a long time just below the eutectic temperature. The copper–silver diagram is given in Figure 9.18.

10 Freezing

Liquids

Liquids have more order than gases but much less than crystals. When a material freezes, its entropy and enthalpy decrease. The enthalpy difference between the liquid and solid states is the latent heat of fusion, ΔH_f, that is released to the surroundings. Similarly, when a metal vapor condenses, the latent heat of vaporization, ΔH_v, is released. For most metals ΔH_v is 20 to 30 times as great as ΔH_f. The difference is because on vaporization all near-neighbor bonds are broken, whereas melting statistically breaks only a fraction of a bond per atom. For coordination numbers of 8 and 12, vaporization breaks four and six bonds per atom. Assuming that both ΔH_f and ΔH_v are proportional to the number of near-neighbor bonds broken, melting must break only a fraction of a bond per atom. The entropy change on melting, $\Delta S_f = \Delta H_f / T_m$, is about 10 MJ/mol K and the entropy change on vaporization, $\Delta S_v = \Delta H_v / T_b$, is about 10 times larger, as shown in Table 10.1.

Most materials contract when they freeze. For most metals the contraction is between 1 and 6%, as shown in Table 10.2 Materials for which packing in the solid is not dense (e.g., Si, Ge, Bi, Ga, and H_2O) actually expand when they solidify.

Homogeneous nucleation

The formation of a tiny sphere of solid in a liquid (Figure 10.1) requires an increase of free energy. The surface energy of the system is increased by $4\pi r^2 \gamma_{LS}$, where γ_{LS} is the energy per area of surface. Because the solid is more stable below the melting point, freezing reduces the free energy by $(4/3)\pi r^3 \Delta G_v$, where ΔG_v is the free energy change per volume transformed. The net change of free energy is

$$\Delta G = 4\pi r^2 \gamma_{LS} - (4/3)\pi r^3 \Delta G_v. \tag{10.1}$$

Table 10.1. Entropy of melting and vaporization of several metals

Element	ΔS^f MJ/mol K	ΔS^v MJ/mol K	$\Delta S_v/\Delta S_f$
Al	11.5	105.	9.15
Bi	20.7	97	4.7
Cd	11.3	95.5	8.4
Ca	7.7	101.5	13.2
Co	9.73	134	13.7
Cu	9.6	104.7	10.9
Ga	18.4	101.2	5.5
Au	9.25	106.9	11.6
Fe	7.71	124.6	16.4
Pb	7.93	96.7	12.2
Li	6.61	97.5	14.7
Mg	9.69	93.3	9.6
Hg	10.1	86.9	8.6
Mo	9.0	84.3	9.4
Re	9.59	103	10.7
Rb	6.96	79	11.3
Ag	11.7	116.5	10.0
Na	7.01	77.2	11.0
Ta	8.85	133.8	15.1
Sn	14.0	93.6	6.7

Source: Data from J. H. Hollomon and D. Turnbull, in *Solidification of Metals and Alloys* (New York: AIME, 1951), p. 13.

As the radius increases, the energy of the system (Figure 10.2) initially increases. However, once a critical radius, r^*, is reached the free energy decreases with further growth so the particle can grow spontaneously. An activation energy, ΔG^*, is required to reach this critical radius. The critical radius and the critical activation energy can be found by differentiating Equation (10.1) and setting $d\Delta G/dr = 0$:

$$d\Delta G/dr = 8\pi r \gamma_{LS} - 4\pi r^2 \Delta G_v = 0 \qquad (10.2)$$

$$r^* = 2\gamma_{LS}/\Delta G_v. \qquad (10.3)$$

Table 10.2. Volume change on melting

Metal	Crystal structure	% vol. change on melting	Metal	Crystal structure	% vol. change on melting
Li	bcc	1.65	Mg	hcp	4.1
Na	bcc	2.2	Zn	hcp	4.2
K	bcc	2.55	Cd	hcp	4.7
Rb	bcc	2.5	Sn	bct	2.8
Cs	bcc	2.6	Hg	rhomb	−1.6
Fe	bcc	3.4	Bi	rhomb	−3.35
Pb	fcc	3.5	Si	dia cub	−12.
Nb	bcc	0.9	Ge	dia cub	−12.
Al	fcc	6.0	water	hex	−8.3

FREEZING

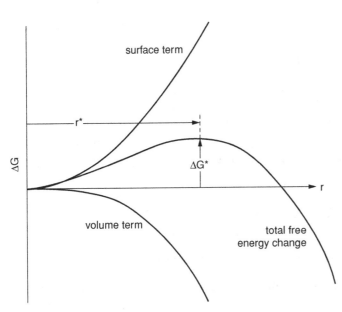

10.1. Spherical embryo of solid forming in a liquid.

10.2. Free energy change during nucleation. The change of free energy, ΔG, increases with embryo size up to a critical radius, r^*. The critical free energy for nucleation is ΔG^*.

Substituting r^* into Equation (10.1),

$$\Delta G^* = (16/3)\pi \gamma_{LS}^3 / \Delta G_v^2. \tag{10.4}$$

Both ΔG_v and ΔG^* become increasingly negative as the temperature is lowered below the melting point:

$$\Delta G_v = \Delta H_v - T \Delta S_v. \tag{10.5}$$

The values of ΔH_v and ΔS_v are almost independent of temperature. For freezing at the equilibrium melting temperature, T_m, $\Delta G_v = 0$ so $\Delta H_v = T_m \Delta S_v$. Below the melting point,

$$\Delta G_v = (T_m - T)\Delta S_v = (\Delta T / T_m)\Delta H_v. \tag{10.6}$$

Table 10.3. Data on subcooling of small droplets

Material	T_m (K)	$\Delta S_f = H_f/T_m$ MJ/(mole K)	ΔT_{max} (K)	$\Delta T_{max}/T_m$
Hg	234	10.0	46	0.197
Sn	506	14.2	110	0.218
Pb	601	8.5	80	0.133
Al	933	11.3	130	0.140
Ag	1234	9.15	227	0.184
Cu	1356	9.6	236	0.174
Ni	1725	10.2	319	0.185
Fe	1803	8.2	295	0.164
Pt	1828	9.4	332	0.182
Water	273	22.1	39	0.143

Source: From J. H. Hollomon and D. Turnbull, in *Solidification of Metals and Alloys* (New York: AIME, 1951).

Substituting into Equation (10.4),

$$\Delta G^* = (16/3)\pi \gamma_{LS}^3/(\Delta H_v \Delta T/T_m)^2 \tag{10.7}$$

so ΔG^* decreases with greater supercooling.

As with other thermally activated processes, the rate of nucleation, $\dot{N}$, can be expressed by an Arrhenius equation with ΔG^* as the activation energy,

$$\dot{N} = \dot{N}_O \exp(-\Delta G^*/kT), \tag{10.8}$$

where $\dot{N}_O$ is a constant and k is Boltzmann's constant.

Substituting Equation (10.7) into Equation (10.8),

$$\dot{N} = \dot{N}_O \exp\{-(16/3)\pi \gamma_{LS}^3/[(\Delta T/T_m)(\Delta H_v)^2 kT]\}. \tag{10.9}$$

Equation (10.9) predicts that the nucleation rate is extremely temperature dependent.

The value of $\dot{N}_O$ in Equation (10.9) has been estimated to be about 10^{39} nuclei/m^{3}s for most metals. This leads to the prediction that a very large subcooling is necessary to produce any nuclei in any reasonable time. For copper at subcoolings of $\Delta T = 100\,°C$, Equation (10.9) predicts that $\dot{N} = 10^{39} \exp[-5.58 \times 10^{-18}/(13 \times 10^{-24} \times 1266)] = 1 \times 10^{-147}$ nuclei/m^{3}s. At this rate of nucleation it would take 3×10^{138} centuries to form one nucleus in a cubic meter of liquid. Undercoolings of $\Delta T \approx 0.18\,T_m$ have been reported for many liquids. See Table 10.3.

Heterogeneous nucleation

Such large undercoolings are normally not observed in metals. Undercoolings are usually so small that they are not noticed. The reason for the difference between the theory and practice is that the theory assumes nucleation occurs homogeneously (i.e., randomly throughout the liquid), whereas nuclei usually form on preexisting solid surfaces.

FREEZING

The importance of special nucleation sites becomes apparent if one watches the formation of gas bubbles in a carbonated beverage. Streams of bubbles rise from certain spots. These are usually small cracks in the glass or dirt particles on a liquid–glass surface.

When nucleation of a solid, S, occurs on a preexisting solid surface, Q, the area between the solid, Q, and the liquid, L, is reduced. However, a new surface is created between S and Q. See Figure 10.3. The net effect is a reduction of the activation energy for nucleation,

$$\Delta G_{hetero*} = \Delta G_{homo*}(2 + \cos\theta)(1 - \cos\theta)^2/4, \qquad (10.10)$$

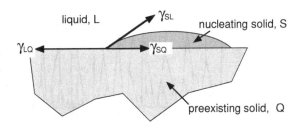

10.3. Heterogeneous nucleation on a preexisting surface, Q. As the new surfaces, SL and SQ, are formed, the surface LQ is lost.

where $\Delta G_{hetero*}$ and ΔG_{homo*} are the activation energies for such heterogeneous and homogeneous nucleation, respectively ($\Delta G_{homo*} = \Delta G^*$ in Equation (10.4)). The wetting angle, θ, is given by

$$\cos\theta = (\gamma_{LQ} - \gamma_{SQ})/\gamma_{SL}. \qquad (10.11)$$

If θ is low, nucleation on the surface is energetically favorable. The preexisting solid surface of most importance in casting is usually the mold wall where the temperature is the lowest. Sometimes nucleating agents with low $-\gamma_{SQ}$ are added to castings to refine the grain size. Crevices or cracks in the mold wall offer special nucleation sites, because they further lower the activation energy.

Growth

Once a solid metal or material with simple molecules (e.g., water) has been nucleated, the freezing rate (velocity of the liquid–solid interface) is controlled almost entirely by the rate of heat removal. The temperature of the solid–liquid interface remains very near the equilibrium freezing temperature. Even if there is substantial undercooling before nucleation, the temperature will rise rapidly back to T_m as freezing occurs because the latent heat, H_f, is large and its release will heat the undercooled liquid.

However, in materials consisting of large molecules (e.g., polymers) or covalently bonded liquids (e.g., siliceous materials), the growth rate may be controlled by the rate at which the molecules can assemble into crystalline form. In this case crystallization may be suppressed with the resulting formation of glass (see Chapter 15).

Grain structure of castings

Many grains are nucleated when a metal is cast against a cold mold. The orientations of these grains are random, but as they grow into the liquid, certain orientations grow slightly ahead of the others and gradually squeeze out the slower growing orientations. This results in columnar structure of the fastest growing orientation, as illustrated in Figure 10.4. In fcc and bcc metals, the <100> direction is the axis of the columnar grains. With hcp metals and water, the columnar axes are perpendicular to the c axis. With continued growth, the degree of alignment increases.

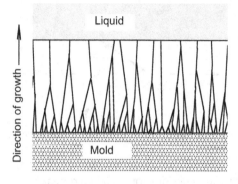

10.4. Columnar grains with axes parallel to the direction of heat flow during freezing. As freezing progresses, the more favorably oriented and faster growing crystals cut off the less favorably oriented, slower growing crystals.

During the freezing of pure materials, the liquid–solid interface is normally planar and crystals grow by the advance of this interface. In contrast, alloys usually freeze by dendritic growth. (The word "dendrite" comes from the Greek, *dendrites*, meaning treelike.) The basic features of dendritic growth are needle-shaped crystals that grow into the liquid and thicken. Usually there are side arms (secondary arms) and sometimes there are tertiary arms (Figure 10.5). The primary arms and the secondary and tertiary arms are crystallographically oriented with <100> being the direction in cubic metals. The reason for dendritic growth will be taken up later. The final columnar structure results from parallel growth of different colonies of dendrites and the gradual lateral growth between them. Whether columnar grains form from plane-front growth as in pure metals or by dendritic growth as in alloys, the final shape and orientations are the same.

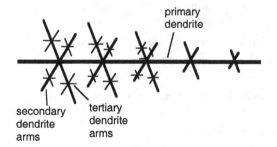

10.5. Schematic drawing of a dendrite with secondary and tertiary arms.

FREEZING

Segregation during freezing

Solute segregation occurs during the freezing of an alloy. The first solid to form is much purer than the overall composition of the alloy (Figure 10.6). As solidification continues, the newly formed solid contains increasing amounts of B. Diffusion in the solid is much too slow to eliminate such concentration gradients. There are several models for predicting solute segregation during freezing, based on the assumption that there is local equilibrium at the liquid–solid interface. If perfect mixing occurred in both the liquid and the solid, the composition of each phase would be that given by the phase diagram.

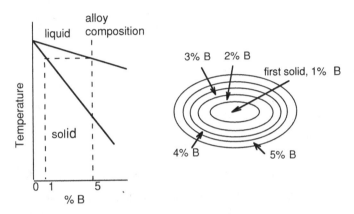

10.6. Schematic drawing showing a binary phase diagram (left) and contours of concentration in a region that has frozen (right).

In analyzing the segregation, Scheil[*] assumed that there is no mixing in the solid and that mixing is perfect in the liquid. The segregation that occurs when an alloy freezes can be modeled by freezing in a horizontal boat (Figure 10.7A). The fraction solid is $f_s = x/L$, where x denotes the position of the solid–liquid interface and L is the mold length, and the fraction liquid is $f_L = 1 - f_s = (L - x)/L$. Figure 10.7B shows the relevant portion of the phase diagram. The composition profile at some time during freezing is illustrated in Figure 10.7C. c_L and c_S are the compositions of the liquid and the solid at the interface, expressed as weight of B per volume. However, they can be expressed as either weight percentage or atom percentage if the density differences are neglected.

As the solid–liquid interface advances a distance, dx, the amount of solute rejected by the solid is $(c_L - c_S)dx$. This solute enriches the liquid composition by dc_L. A mass balance gives

$$(c_L - c_S)dx = (L - x)dc_L. \tag{10.12}$$

The liquidus and solidus can usually be approximated by straight lines. Then at all temperatures,

$$c_S = kc_L, \tag{10.13}$$

[*] E. Scheil, *Z. Metallkunde* 34 (1942): 70.

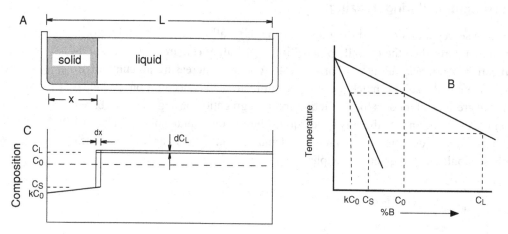

10.7. Plane-front solidification in a horizontal mold.

where k is the *distribution coefficient*. Substituting $c_S = kc_L$, $f_S = x/L$, and $dx = L df_S$, $c_L = c_0$ at $f_S = 0$ and c_L at f_S, so $kc_L \int_0^{f_s} df_S/(1-f_S) = 1/(1-k) \int_{c_0}^{c_L} dc_L/c_L$. $[1/(1-k)] \ln(c_L/c_0) = -\ln(1-f_S)$ or

$$c_L = c_0(1-f_S)^{-(1-k)} \quad \text{and} \quad c_S = kc_0(1-f_S)^{-(1-k)}. \tag{10.14}$$

Equation (10.14) is called the Scheil equation. Figure 10.8 shows its predictions of how the composition of the ingot changes with f_S for $c_0 = 5\%$ and $k = 1/5$. It should be noted that Equation (10.14) is valid as long as no diffusion occurs in the solid and there is perfect mixing in the liquid. It applies for even dendritic growth.

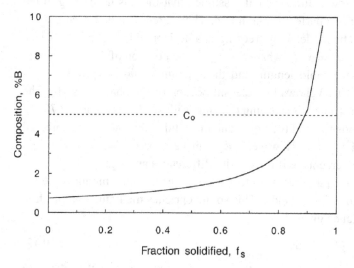

10.8. Segregation during freezing for an alloy containing 5% B and having a distribution coefficient of $k = 0.2$.

For a given alloy, the nonequilibrium solidus (average solid composition) can be calculated as a function of the temperature. The average composition of the solid,

FREEZING

$\bar{c}_s$, can be found from solving the simple mass balance $c_0 = \bar{c}_s f_s + c_L(1 - f_s)$, $\bar{c}_s = [c_0 - c_L(1 - f_s)]/f_s$. Substituting $f_s = 1 - (c_L/c_0)^{-1/(1-k)}$ from the Scheil equation,

$$c_S = [c_0 - c_L(c_L/c_0)^{-1/(1-k)}]/[1 - (c_L/c_0)^{-1/(1-k)}]. \tag{10.15}$$

Figure 10.9 shows the nonequilibrium solidus calculated for an alloy containing 1% B and A melting at 1660 °C. The liquidus temperature is approximated by $T_L = 660 - 40c_L$ or $c_L = (660 - T)/40$ and $c_S = (660 - T)/200$, so the distribution coefficient is 0.20.

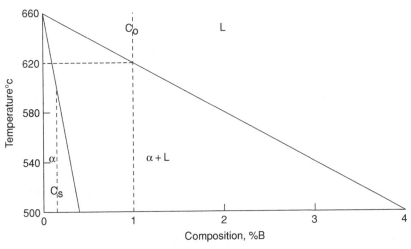

10.9. Plot of the nonequilibrium solidus for an alloy of 1% B, superimposed on the phase diagram.

Zone refining

Materials can be purified by directional solidification by cropping off and saving the first end to freeze. A material may be further purified by combining first ends from several ingots and repeating the process. A simpler, continuous process called *zone melting* was proposed by W. Pfann. This involves repeatedly passing molten zones through the ingot, as shown in Figure 10.10. Although the first pass in zone refining produces less purification than directional solidification, further purification can be achieved by passing additional zones through the material. Figure 10.11 shows calculations of the purification by successive passes.

10.10. Zone refining. As a liquid zone is passed slowly from left to right, it collects impurities.

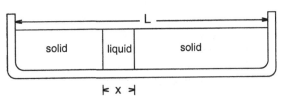

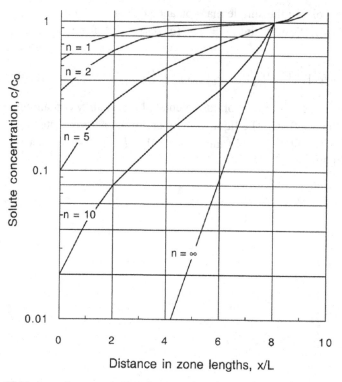

10.11. Impurity concentration after a number of passes. Adapted from W. G. Pfann, *Zone Melting* (New York: Wiley, 1958). Reprinted with permission of John Wiley & Sons, Inc.

The actual purification is somewhat less than that predicted by Equation (10.15) and by Figure 10.11 because mixing is not perfect in the liquid. A thin boundary layer of impurity forms in the liquid just ahead of the interface. See Figure 10.12. The formation of a boundary layer causes the concentration of the impurity in the liquid at the liquid–solid interface to be greater than for perfect mixing. Therefore, there is less purification. Zone refining is most efficient in relatively pure materials, where the boundary layer builds up.

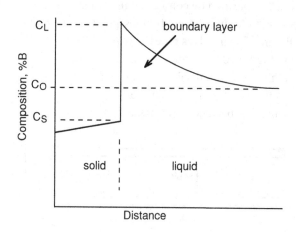

10.12. Boundary layer ahead of liquid–solid interface.

FREEZING

Steady state

A steady-state condition is reached when the boundary layer becomes great enough that the solid forming has the same composition, c_0, as the liquid beyond the boundary layer. The liquid composition at the interface of the boundary layer is c_0/k. This is illustrated in Figure 10.13.

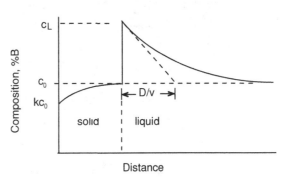

10.13. Steady-state freezing. The interface composition in the liquid is c_0/k, so the solid that forms has the composition c_0.

At steady state, the thickness, t, of the boundary layer can be approximated as

$$t = D/v, \tag{10.16}$$

where D is the diffusivity of the solute in the liquid and v is the velocity of the solid–liquid interface.

Dendritic growth

The formation of a boundary layer may lead to a breakdown of plane-front growth and dendritic growth. Figure 10.14 shows dendrites forming during solidification of a solution of polymers. The variation of the composition in the liquid of the boundary layer (Figure 10.15A) causes a variation in the local liquidus temperatures in accordance with the phase diagram (Figure 10.15B). Both the liquidus temperature and the actual temperature are plotted in Figure 10.15C. The actual temperature at the interface equals the liquidus temperature. The actual temperature just ahead of the interface is lower than the liquidus temperature for the local composition. This condition is called *constitutional supercooling* because it results from compositional (constitutional) variations. This situation is unstable. If the solid anywhere in the interface happens to extend slightly ahead of the other places, it will freeze faster and grow rapidly into the undercooled liquid, leading to dendritic growth.

Figure 10.16 shows that the critical thermal gradient for prevention of dendritic growth is

$$(dT/dx)_{\text{crit}} = (T_1 - T_3)/(D/v), \tag{10.17}$$

where T_1 and T_3 are the liquidus temperatures of the compositions c_0 and c_0/k.

10.14. Dendrites forming during the freezing of a transparent polymer solution. Reprinted with permission of ASM International® from K. A. Jackson, *Solidification* (Materials Park, OH: ASM, 1971), p. 121. All rights reserved. www.asminternational.org.

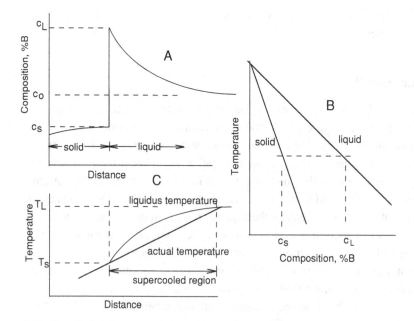

10.15. Constitutional supercooling resulting from boundary layer formation.

Sometimes a distinction is made between *cellular growth* and dendritic growth. In cellular growth the primary arms extend into the liquid but there are no secondary dendrite arms. Figure 10.17 illustrates this. In either case, however, primary arms extend in the direction of heat flow. The tendency to cellular and

FREEZING

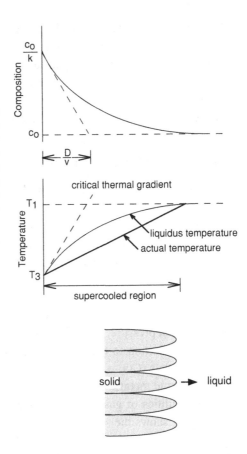

10.16. (A) The composition profile near the liquid–solid interface for steady-state freezing. (B) The temperature in the liquid near the interface, T_L, and the equilibrium liquidus temperature, T_e, corresponding to the local composition. The region where $T_L < T_e$ is supercooled so dendrites can form. Dendrites cannot form if the actual thermal gradient is greater than the critical gradient.

10.17. Cellular growth.

dendritic growth increases as the thermal gradient, G, decreases and as the growth velocity, v, increases.

The length of dendrites, L, can be estimated from knowledge of the thermal gradient and the phase diagram. For a given alloy, the temperature at the tips of the dendrites is the liquidus temperature of the alloy, T_L, and the temperature at the base of the dendrites is the solidus temperature, T_S. While this may be lower than the equilibrium solidus of the alloy, the separation of the liquidus and solidus temperatures gives an approximate indication of the relative tendency to form long or short dendrites:

$$L = (T_L - T_S)/(dT/dx). \tag{10.18}$$

The spacing, λ, between secondary dendrite arms has been shown to increase with solidification time, t_f,

$$\lambda = k(dT/dt)^p, \tag{10.19}$$

where the exponent p is about 1/3. With dendritic growth the segregation is almost entirely interdendritic rather than macroscopic. The distances between the concentration minima are the distances between dendrite arms.

Dendritic growth also affects the nature of porosity in castings. Without dendritic growth, shrinkage results in large cavities in the last regions to freeze. If the dendrites are so long that liquid cannot flow easily through the interdendritic channels to compensate for the shrinkage, the shrinkage will occur interdendritically on a microscopic scale. With such interdendritic shrinkage, the macroscopic shrinkage will be absent or greatly reduced.

Gas solubility and gas porosity

Most gases dissolve monatomically in liquid metals. For example, the solution reactions may be written as $H_2 \to 2\underline{H}$, $N_2 \to 2\underline{N}$, $O_2 \to 2\underline{O}$, where the underlining signifies the element is in solution. Sievert's law for diatomic gases is an application of the mass action principle. It states that the solubility is proportional to the square root of the partial pressure of the gas. For example,

$$\underline{H} = k(P_{H_2}), \tag{10.20}$$

where $\underline{H}$ is the concentration of the dissolved hydrogen that is in equilibrium with the partial pressure, P_{H_2}, of hydrogen gas and k is a temperature-dependent constant. In addition to diatomic gases, carbon monoxide and water vapor are soluble in metals ($CO \to \underline{C} + \underline{O}$) and $H_2O \to 2\underline{H} + \underline{O}$). Hydrogen is soluble in almost all liquid metals.

The solubilities of gases in solid metals are much lower than liquid metals. Figure 10.18 shows the solubility of hydrogen in copper and copper–aluminum alloys. Because of the lower solubility in the solid, gas bubbles are released at the liquid–solid interface as the metal freezes. With long dendrites the gas bubbles are trapped and the result is gas porosity.

Growth of single crystals

Single crystals may be grown by directional solidification. With the Bridgman technique, a mold is slowly removed from a furnace. Freezing starts at one end and slowly progresses to the other. Because freezing starts at a point, only one crystal is nucleated. The Czochralski method involves lowering a seed crystal into melt so that it partially melts and then withdrawing it slowly upward. The growing crystal is rotated about a vertical axis to help stir the liquid. The lack of a mold eliminates contamination from mold walls but makes it impossible to control the exterior shapes of the crystals. This method is used to grow silicon crystals for the semiconductor industry.

Eutectic solidification

As a eutectic front advances during solidification, the solutes must partition between the two phases, as suggested in Figure 10.19. With increased rates of solidification, there is less time for diffusion so the eutectic structures are finer.

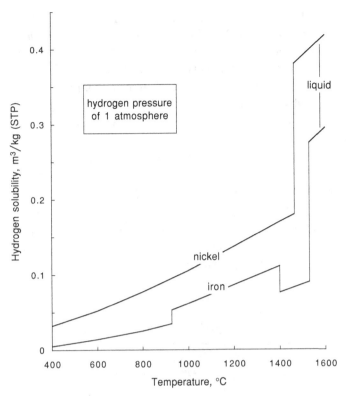

10.18. The solubility of hydrogen in iron and nickel as a function of temperature. From A. Guy, *Elements of Physical Metallurgy* (Reading, MA: Addison-Wesley, 1959), p. 210.

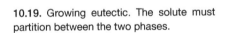

10.19. Growing eutectic. The solute must partition between the two phases.

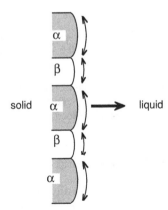

Eutectic reactions, liquid → $\alpha + \beta$, can result in several geometric configurations of α and β. When the volumes of both phases are nearly equal, the most common morphology is lamellar. This is true of the Cu–Ag and Pb–Sn eutectics. If the amount of one phase is much less than the other, the eutectic is likely to be rods of one phase surrounded by the other phase (e.g., NiAl–Cr, TaC–Ni eutectics.) If the volume fraction of one phase is very low, that phase may form as isolated islands (e.g., graphite in Fe, Si in Si–Al alloys.) Whether the morphology is in the form of platelets, rods, or isolated spheres depends on the volume fraction of the two phases. The most likely morphology is the one that minimizes the total interphase area and depends on the volume fraction, f, of the minor phase.

For spheres in a simple cubic array, the volume fraction is $f = (4/3)\pi(r/\lambda)^3$, where λ is the separation distance and the surface area per volume, A_v, is $4\pi r^2/\lambda^3$. Combining, $A_v = (4\pi)^{1/3}(3f)^{2/3}/\lambda$.

For a square array of rods, $f = \pi(r/\lambda)^2$ and $A_v = 2\pi r/\lambda^2$. Combining, $A_v = 2(f\pi)^{1/2}/\lambda$.

For parallel platelets, the surface area per volume, A_v, is $2/\lambda$, regardless of f. Figure 10.20 is a plot of $A_v/(A_v)_{\text{parallel plates}}$ for the three geometries as a function of f. According to this simple analysis, parallel plates have the least area for $f \geq 1/\pi = 31.8\%$, rods the least area for $4\pi/81 (= 15.5\%) \leq f \leq 1/\pi (31.8\%)$ and isolated spheres for $f \leq 4\pi/81 = 15.5\%$.

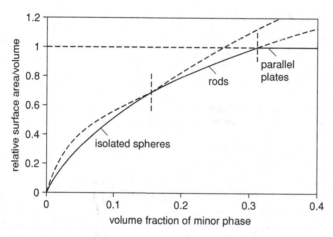

10.20. Relative amount of interphase area in eutectics composed of platelets, rods, and spheres. The morphology with the lowest interphase area has the least energy. From W. F. Hosford, *Physical Metallurgy* (Boca Raton, FL: CRC Press, 2005).

Peritectic freezing

The freezing of an alloy of the peritectic composition, c_0, is schematically illustrated in Figure 10.21. During the reaction $\alpha + \text{liquid} \rightarrow \beta$, the β forms on the surface of preexisting α where there is contact between the α phase and the liquid. The film of β prevents direct contact between the liquid and α. Further reaction can occur only by diffusion of A or B atoms through the β, so as the film thickens, the reaction becomes extremely slow. Usually peritectic reactions do not go to completion. Microstructures usually contain α phase, even though the phase diagram predicts it should not exist. The term *surrounding* is used to describe this phenomenon.

FREEZING

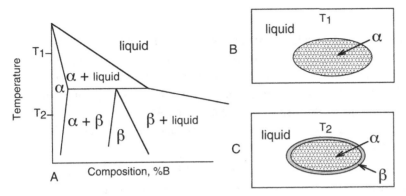

10.21. (A) Portion of a phase diagram with a peritectic reaction. (B) Particle of α just before the peritectic reaction. (C) The same region as β begins to form between the α and the liquid.

NOTES OF INTEREST

1. Refrigerator ice cubes have tiny hollow tubular channels oriented in the direction of freezing. Because the solubility of air in ice is much less than in liquid water, air is released as the ice forms. Very careful examination of these "ice worms" will reveal that their diameter changes periodically, as shown in Figure 10.22. The cause of this periodicity became a subject of some interest at one of the leading industrial laboratories until someone observed that the periodicity was related to the on–off cycle of the refrigerator. When the temperature was low the channels were wider because the faster freezing occurred more rapidly, allowing less time for diffusion of air to the surface.

10.22. "Worm" in an ice cube.

2. Crevice nucleation can be observed in a glass of carbonated beverage. A series of bubbles can be seen rising from the same place. The source of these bubbles is a favored nucleation site, probably a minute scratch in the bottom of the glass.

REFERENCES

W. F. Hosford. *Physical Metallurgy*. Boca Raton, FL: CRC Press, 2005.
K. A. Jackson. *Solidification*. Materials Park, OH: ASM, 1971.
W. Kurz and D. J. Fisher. *Fundamentals of Solidification*. Üticon-Zürich: Trans Tech Publications, 1984.
D. A. Porter and K. E. Easterling. *Phase Transformations in Metals & Alloys*. London: Chapman & Hall, 1981.

PROBLEMS

1. The rate at which metals freeze is controlled by the rate at which heat can be extracted. Consider the freezing of copper. If the liquid–solid interface advances at 1 mm/s, what is the thermal gradient (°C/mm) in the solid?
   Data for aluminum: melting point = 1085 °C, specific heat = 400 J/(kg – °C), heat of fusion = 205 J/kg, atomic wt. = 63.5, density = 8.9 Mg/m³, thermal conductivity = 160 (W/m – °C), and coefficient of linear expansion = 17×10^{-6}/°C.

2. An ingot of Al–4% Cu is directionally solidified. Assume that there is no diffusion in the solid and that there is perfect mixing in the liquid. Pure aluminum melts at 660 °C. At the eutectic temperature of 548 °C, the liquid composition is 33.2% Cu and the solid composition is 5.35% Cu. Assume that the liquidus and solidus are straight lines.
   A. Find the distribution coefficient expressed as $k = c_S/c_L$, where c_S and c_L are expressed as % Cu.
   B. Calculate the composition of the liquid when the solidification is 60% complete.
   C. What is the average composition of the solid, $\bar{c}_S$, at this point. (Make sure that $0.60\bar{c}_S + 0.40c_L = 4\%$.)
   D. What is the liquid–solid interface temperature at this point?
   E. How much eutectic will be formed?

3. Consider the freezing of an aluminum alloy containing 0.002% copper.
   A. What would be the composition of the first solid to freeze?
   B. What would be the average composition of the first half to freeze?

4. Consider the steady-state freezing of an aluminum alloy containing 0.55% Cu. In steady-state freezing the boundary layer is such that the solid freezing has the same composition as the alloy. Assume that the liquid–solid interface moves at a rate of 80 µm/s. The diffusion coefficient of copper in liquid aluminum is 3×10^{-9} m²/s.
   A. What is the interface temperature?
   B. What is the thickness of the boundary layer?
   C. What temperature gradient would be required to maintain plane-front growth?

5. At the melting point of aluminum and one atmosphere partial pressure of hydrogen, the equilibrium solubility of hydrogen is 7×10^{-3} cm³/g of Al in the liquid and 4×10^{-4} cm³/g of Al in the solid. The solubilities, 4×10^{-4} cm³/g and 7×10^{-3} cm²/g, are expressed as the volumes measured at 20 °C and 1 atmosphere (STP), *not* the volumes at the melting point.
   A. Calculate the equilibrium solubilities in the liquid and solid at 0.2 atmosphere H_2. Express your answer in STP.
   B. What volumes of H_2 would be liberated during the freezing per volume of aluminum, if the partial pressure of H_2 were 0.2 atmospheres? (The

H$_2$ is liberated at a total pressure of 1 atmosphere and at the melting point of aluminum.) Assume the perfect gas law. Your answer should equal the percentage of gas porosity if the gas is trapped interdendritically.

6. Consider an aluminum–rich binary aluminum–silicon alloy. The melting temperature of aluminum is 660 °C, and the eutectic is at 577 °C and 12.6 wt% Si. The maximum solubility of silicon in aluminum is 1.65% Si at 577 °C. The liquidus and solidus can be approximated by straight lines. The diffusivity of silicon in liquid aluminum is 8×10^{-8} m^2/s. Freezing occurs at a rate of 10 μm/s.
   A. For an alloy of 0.05% Si, what is the interface temperature for steady-state freezing?
   B. Find the thickness of the boundary layer.
   C. What temperature gradient is necessary to maintain plane-front growth?
   D. Repeat A, B, and C for an alloy containing 1% silicon.

7. Predict the morphology of each of the eutectics listed below. The compositions are from phase diagrams in the *Metals Handbook*, vol. 8, 8th ed. (1973). Some of the densities are estimates.

System	Phase	Composition	Density
Sn/Pb	α Sn	97.5% Sn	7.3 Mg/m^3
	Eutectic	61.9% Sn	
	βPb	19% Sn	10.6
Cu/Cu$_2$O	αCu	0.036% O	8.9
	Eutectic	0.39% O	
	Cu$_2$O	11.3% O	8.7
Bi/Pb	BiPb$_2$	42% Bi	11.4
	Eutectic	56% Bi	
	Bi	100% Bi	9.8

Source: Reprinted with permission of ASM International®. All rights reserved. www.asminternational.org.

8. The melting point of pure aluminum is 660 °C and aluminum and silicon form a eutectic, the eutectic temperature is 577 °C, the eutectic composition is 12% Si, and the maximum solubility of of silicon in solid aluminum is 1.65%. Assume the phase diagram consists of straight lines. If aluminum containing 0.15 wt% Si were solidified, what would be the composition of the first solid to form?

9. Some solutes raise the melting temperature, causing both the liquidus and solidus to increase with additional solute. In this case the distribution coefficient is $k > 1$.
   A. Is the Scheil equation (10.14) still valid?
   B. Describe qualitatively how having $k > 1$ affects the segregation.

11 Phase Transformations

Solid-state reactions may be classified by whether or not they require both nucleation and diffusional growth: Eutectoid and precipitation reactions require nucleation and growth by long-range diffusion. Examples include pearlite formation in iron–carbon alloys and precipitation of the θ phase in aluminum–copper alloys. In massive transformations, such as transformation of the bcc to fcc structures in pure iron, there is no change of composition so only nucleation and local readjustment of atom positions are required. Martensitic reactions occur by nucleation and shear without any composition change so diffusion is not required. An example is the formation of iron–carbon martensite on quenching austenite. Finally, there is no nucleation stage in spinodal transformations that occur on cooling into a miscibility gap. However, diffusion is required.

Nucleation in the solid state

Nucleation of a new phase in the solid state is more complicated than that of nucleation in freezing. Volume difference between the new and old phases causes an elastic misfit term that increases ΔG. Destruction of existing grain boundaries reduces ΔG. An expression for the free energy change during nucleation of β in a matrix of α is

$$\Delta G = \gamma_{\alpha\beta} \Delta A_{\alpha\beta} - \gamma_{\alpha gb} \Delta A_{\alpha gb} + \Delta G_v \Delta V_\beta + \text{an elastic strain energy term,}$$

(11.1)

where $\gamma_{\alpha\beta}$ and $\Delta A_{\alpha\beta}$ are the α–β interfacial energy and change of α–β interfacial area and $\gamma_{\alpha gb}$ and $\Delta A_{\alpha gb}$ are the α grain boundary energy and the change of α grain boundary area.

In general, the nuclei are not spherical. There are several reasons for this. One is that the α–β surface energy term, $\gamma_{\alpha\beta}$, depends on the orientations of the α and β phases. Another reason is that when nucleation occurs on α–α grain boundaries, the angle of contact, θ, between the α and β phases depends on the ratio of $\gamma_{\alpha\beta}/\gamma_{\alpha gb}$. A third reason is that the elastic strain energy is minimized if the precipitating phase is lenticular.

PHASE TRANSFORMATIONS

Grain boundaries, grain edges, and grain corners are special sites for nucleation (Figure 11.1) because the formation of a new phase eliminates some grain boundaries in the old phase. Figure 11.2 shows the dependence on the critical free energy for nucleation at these special sites on the wetting angle, θ. The equilibrium wetting angle, θ, depends on the ratio of the energy, $\gamma_{\alpha gb}$, of α–α grain boundaries and the interfacial energy, $\gamma_{\alpha\beta}$, between α and β grains, as shown in Figure 11.3. A force balance gives

$$2\gamma_{\alpha\beta}\cos\theta = \gamma_{\alpha\beta}. \tag{11.2}$$

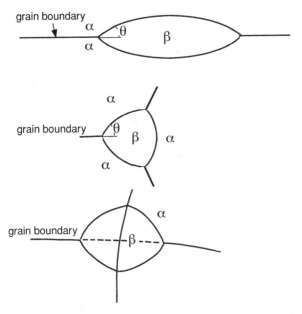

11.1. Special nucleation sites on grain boundaries, along grain edges, and at grain corners.

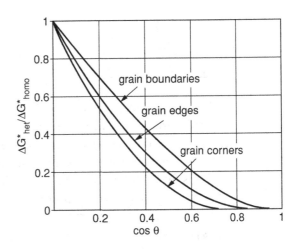

11.2. Relative energy of a nucleus at special sites as a function of the equilibrium wetting angle. Adapted from J. W. Cahn, *Acta Met.* 4 (1956): 456.

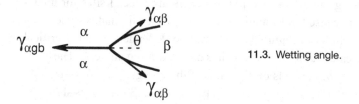

11.3. Wetting angle.

Eutectoid transformations

The growth rate of eutectoids is controlled by the rate at which the components can separate into two different phases. Figure 11.4 illustrates schematically the reaction $\gamma \to \alpha + \beta$, where α is rich in component A, and β is rich in component B. Component A must diffuse away from the advancing β and B away from the advancing α. This diffusion occurs in the α–γ and β–γ boundaries. In the case of a pure binary alloy, only A and B need to diffuse, so the rate of growth will be determined by the diffusivities of these components. In systems with more than two components, the other components partition between the α and β phases, so they must diffuse as well. Alloying elements in steels diffuse much slower than carbon, so the rate of growth of pearlite is controlled by the diffusion of the alloying elements.

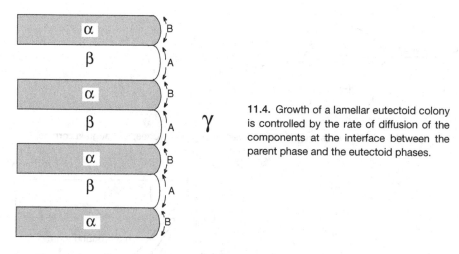

11.4. Growth of a lamellar eutectoid colony is controlled by the rate of diffusion of the components at the interface between the parent phase and the eutectoid phases.

Although the eutectoid structure in Figure 11.4 consists of parallel platelets, other geometric arrangements are possible. The most likely arrangement is the one that minimizes the diffusion distances. The spacing of platelets or rods depends on the temperature at which the reaction occurs. With greater supercooling, the spacing is finer. This compensates somewhat for the slower diffusion rates. Figure 11.5 shows the decrease of pearlite spacing as the transformation temperature is lowered below the equilibrium eutectoid temperature.

It is possible to cool steels so that pearlite is formed at a temperature, T, below the equilibrium eutectoid temperature, T_e. In this case, the platelet spacing, λ,

PHASE TRANSFORMATIONS

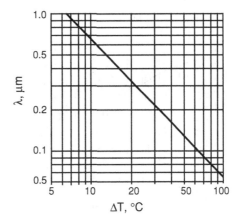

11.5. With lower transformation temperatures, the lamellae spacing of pearlite is finer. Data is from D. D. Pearson and J. D. Verhoeven, *Met. Trans., A* 15A (1984): 1037.

is inversely proportional to the undercooling, $\Delta T = T_e - T$, as shown in Figure 11.5:

$$\lambda = A/\Delta T, \tag{11.3}$$

where A is a composition-dependent constant.

A simple analysis can be made of the temperature dependence of the growth rate of pearlite. The lamellae spacing has two effects on the growth rate. The diffusion gradient, dc/dx, depends on λ and the flux required for a given rate of growth is proportional to λ. According to Fick's first law, the flux of carbon to the growing carbide is

$$J = -D dc/dx = D(c_c - c_\gamma)/(f_d \lambda), \tag{11.4}$$

where c_c and c_γ are compositions of the carbide and austenite and f_d is the ratio of the effective diffusion distance to λ. The flux must equal the rate that carbon is incorporated into the growing carbide:

$$J = f_c(c_c - c_\gamma) dx/dt, \tag{11.5}$$

where f_c is the volume fraction carbide. Combining Equations (11.3), (11.4) and (11.5) and substituting $dx/dt = G$,

$$G = BD\Delta T^2, \tag{11.6}$$

where B is a constant. Because $D = D_o \exp[-Q/(RT)]$,

$$G = C\Delta T^2 \exp[-Q/(RT)], \tag{11.7}$$

where C is a new constant.

The temperature dependence caused by the $\exp[-Q/(RT)]$ term is opposite to that of the $(\Delta T)^2$ factor. The net effect is that there is a maximum growth rate near 600 °C, as shown in Figure 11.6. Figure 11.6 also shows that the nucleation rate, N, continues to increase with lower temperatures.

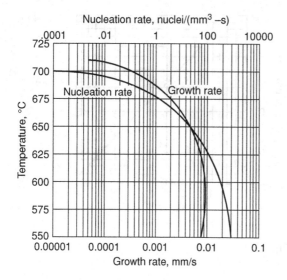

11.6. Temperature dependence of pearlite nucleation and growth rates in a 0.78% C, 0.63% Mn steel of ASTM grain size 5.25. Data from R. F. Mehl and A. Dube, *Phase Transformations in Solids* (New York: Wiley, 1951), 545. Reprinted with permission of John Wiley & Sons, Inc.

The net effect is that the overall transformation rate has a maximum somewhat below 600 °C, as shown in Figure 11.7. The time for transformation is inversely related to the transformation rate. Isothermal transformation diagrams (or TTT curves) are plots of the time required for transformations to occur. Because time is reciprocally related to rate, isothermal transformation diagrams have a C shape, as shown in Figure 11.8.

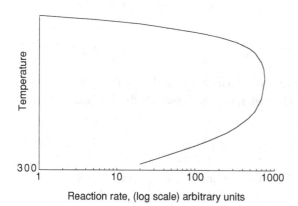

11.7. Schematic figure showing how a transformation rate varies with temperature.

Avrami kinetics

Johnson and Mehl[*] analyzed the kinetics of pearlite formation by assuming that the growth rates, G, in three dimensions and the nucleation rate, N, are constant. With a constant growth rate, the volume, V, of a spherical particle nucleated at a time, τ, at a time, t, is

$$V = (4\pi/3)G^3(t - \tau)^3. \tag{11.8}$$

[*] W. A. Johnson and R. F. Mehl, *Trans. AIME* 135 (1939): 416.

PHASE TRANSFORMATIONS

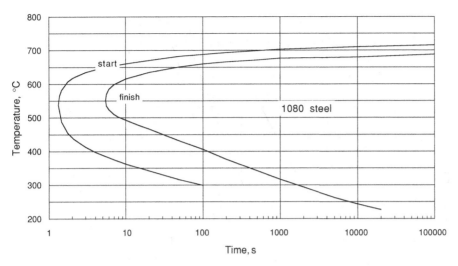

11.8. Isothermal transformation of a 1080 steel. The left-hand line is for the observable start of the reaction and the right-hand line is for the essential completion of the reaction. Data from *Atlas of Isothermal Transformation Diagrams* (Pittsburgh: U.S. Steel, 1951).

The number of nuclei per volume of untransformed material formed in a time increment, $d\tau$, is $Nd\tau$. In the early stage of transformation impingement of transformed material may be neglected. In that case, the fraction transformed is

$$f = (4\pi/3)G^3 N \int (t-\tau)^3 d\tau = (\pi/3)G^3 N t^4. \tag{11.9}$$

However, this is valid only for $f \ll 1$. At longer times the fraction of the volume that is available for growth and nucleation is $(1-f)$. When this term is included, the Johnson–Mehl equation becomes

$$f = 1 - \exp[-(\pi/3)G^3 N t^4]. \tag{11.10}$$

For cases in which the nucleation and the growth rates are not constant or the growth is not three dimensional, Avrami[*] showed that this equation can be generalized to

$$f = 1 - \exp(-kt^n), \tag{11.11}$$

where the exponent n may be less than 4. The effect of the exponent, n, on the transformation is shown in Figure 11.9. The values of k were not adjusted, so the apparent faster transformation rate with lower exponents is not real. The ratio of 90% and 10% completion times is less for the higher value of n.

EXAMPLE 11.1. Find the ratio of the times for a reaction to be 90% complete to the time for it to be 10% complete. Compare that ratio for $n = 4$ and $n = 2$.

SOLUTION: From Equation 11.10, $\ln(1-f) = -kt^n$. Comparing two degrees of completion, $\ln(1-f_2)/\ln(1-f_1) = (t_2/t_1)^n$ so $t_2/t_1 = [\ln(1-f_2)/\ln(1-f_1)]^{1/n}$.

[*] M. Avrami, *J. Chem. Phys.* 7 (1939): 1103.

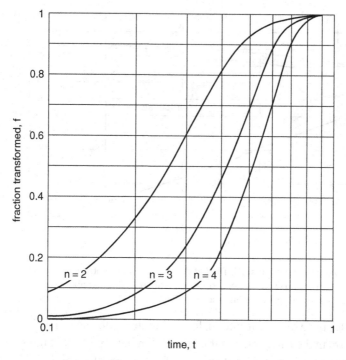

11.9. The effect of the exponent in the Avrami equation on the transformation. On a log(f)-log(t) plot, the slope increases with n.

Substituting $f_2 = 0.9$ and $f_1 = 0.1$, $t_2/t_1 = 21.8^{1/n}$. For $n = 4$, $t_2/t_1 = 2.16$; for $n = 2$, $t_2/t_1 = 4.7$.

The constant, k, which incorporates both G and N, is very temperature sensitive because both G and N depend on temperature. Changes in k shift the curve horizontally on a semilogarithmic plot but do not change its shape. Figure 11.10 shows the rate of transformation for two temperatures, T_1 and T_2, with $n = 4$.

For 50% transformation, $-kt_{(0.5)}{}^n = \ln(0.5)$, where $t_{(0.5)}$ is the time for $f = 50\%$, so

$$k = .69/t_{(0.5)^n}. \tag{11.12}$$

There are several reasons why the Avrami exponent may be less than 4. The nucleation rate may decrease with time because most favorable nucleation sites are used up early. In the extreme, it is possible that all nucleation sites are used up at the very start so nucleation makes no contribution to the exponent. In some cases, the growth rate may decrease with time. This is true for precipitation from solid solution. For precipitation, the rate of growth is inversely proportional to the square root of time. Finally, if growth is in only one or two dimensions instead of three, the contribution of growth will contribute less to the exponent. Table 11.1 lists the contributions to the exponents.

PHASE TRANSFORMATIONS

Table 11.1. Contributions to the Avrami exponent

Growth	Constant G	$G \propto \delta t^{-1/2}$
3-dimensional	3	1.5
2-dimensional	2	1
1-dimensional	1	0.5
Nucleation	Constant N	Site saturation
	1	0

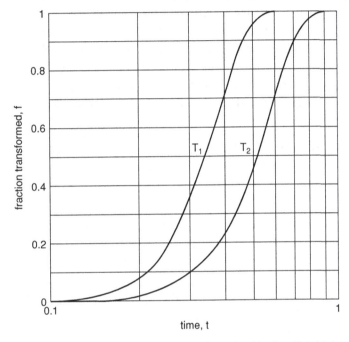

11.10. The effect of temperature on transformation kinetics. T_1 is higher than T_2.

EXAMPLE 11.2 A reaction $\dot{N}$ is 20% complete after 45 s and 85% complete after 1.25 min. Determine the value of n in the Avrami equation.

SOLUTION: Writing the Avrami equation as $-\ln(1-f) = bt^n$ and evaluating at two conditions,

$$\ln(1-f_2)/\ln(1-f_1) = bt_2{}^n bt_1{}^n = (t_2/t_1)^n \quad \text{so}$$
$$n = \ln[\ln(1-f_2)/\ln(1-f_1)]/\ln(t_2/t_1)$$
$$= \ln(\ln 0.8/\ln 0.15)/\ln(45/1.25 \times 60) = 4.2.$$

Many reactions, including recrystallization, can be described by Avrami kinetics.

Growth of precipitates

Diffusion of the solute is necessary for precipitate growth. The diffusional flux depends on both the diffusivity and the concentration gradient, which is proportional to $c_o - c_\alpha$. At low amounts of supercooling, the term $c_o - c_\alpha$ is nearly proportional to the degree of supercooling so the growth rate increases with supercooling. However, at high supercoolings, the temperature dependence of the diffusivity masks this effect and the growth rate decreases. The rate of nucleation increases with supercooling, resulting in a finer dispersion.

At low amounts of supercooling the driving force for nucleation is so small that grain boundary nucleation predominates. Figure 11.11 shows schematically the phase diagram of an alloy in which precipitates can form. If an alloy solution is treated and then quenched to temperature T, particles of β will precipitate out of the α. The α phase at the α–β boundary will have the composition c_α given by the phase diagram, whereas the composition of the α phase remote from the particle will remain c_o. The concentration gradient, dc/dx, in the α-phase allows diffusion of element B to the growing β. The growth rate of β is proportional to the diffusion flux, $J = -Ddc/dx$. The concentration gradient, dc/dx, can be approximated as $(c_o - c_\alpha)/L$, as indicated in Figure 11.12. A mass balance requires that $(c_\beta - c_o)x = L(c_o - c_\alpha)/2$ so $L = 2(c_\beta - c_o)x/(c_o - c_\alpha)$ and

$$J = -Ddc/dx = -D(c_o - c_\alpha)^2/[2(c_\beta - c_o)x]. \qquad (11.13)$$

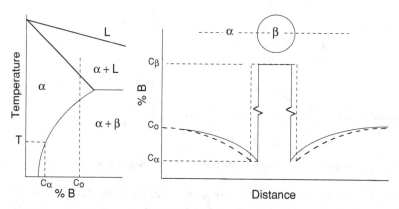

11.11. Phase diagram of A–B system (left) showing the alloy composition, c_{av}, and the compositions of the α and β at the precipitation temperature, T. At the right is the composition profile through a β particle and the neighboring α. Particle growth requires diffusion of B through the α. As the particle grows, the concentration gradient decreases.

The flux, J, is given by $J = (c_\beta - c_o)dx/dt$. Equating, $(c_\beta - c_o)dx/dt = -D(c_o - c_\alpha)^2/[2(c_\beta - c_o)x]$. Rearranging and integrating,

$$\int x\,dx = -D[(c_o - c_\alpha)/(c_\beta - c_o)]^2/2 \int dt \quad \text{or}$$

$$x = [(c_o - c_\alpha)/(c_\beta - c_o)](Dt)^{1/2}. \qquad (11.14)$$

The growth rate $G = dx/dt = (1/2)[(c_o - c_\alpha)/(c_\beta - c_o)](D/t)^{1/2}$.

Thus, the growth rate decreases with time and the particle radius is proportional to $\sqrt{(Dt)}$. The growth rate slows even more when diffusion fields overlap, as shown in Figure 11.13.

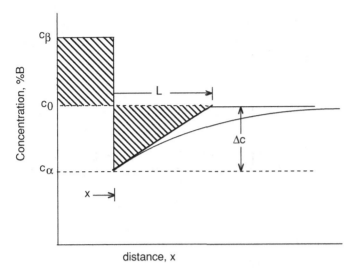

11.12. The concentration gradient near a growing precipitate. The dashed line is a linear approximation to the gradient. The areas $(c_\beta - c_0)x$ and $(1/2)(c_0 - c_\alpha)L$ are equal.

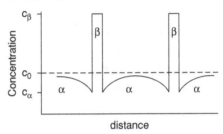

11.13. Overlapping of the diffusion fields of two particles slows the growth. From W. F. Hosford, *Physical Metallurgy* (Boca Raton: CRC Press, 2004), p. 198, figure 10.18.

Transition precipitates

Often a transition precipitate forms before the equilibrium precipitate. Figure 11.14 is a schematic plot of the free energy versus composition curves for such a case. Note that when α is in metastable equilibrium with β' its solubility for B is greater than when α is in equilibrium with β.

Precipitation-free zones

If the precipitation occurs at high temperatures, the degree of supersaturation will be low and the precipitates will grow rapidly. These conditions lead to formation of large precipitates at the grain boundaries. The growth of these precipitates drains the region near the grain boundaries of solute, so a precipitation-free zone will form near the grain boundaries. An example is shown in Figure 11.15. Quenching minimizes the width of the precipitation-free zone. This is illustrated schematically in Figure 11.16.

Ostwald ripening

During precipitation the number of particles initially increases with time as more nuclei are formed. Eventually a maximum is reached and then the number of

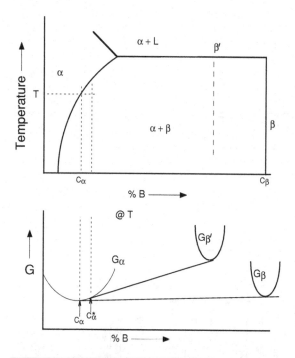

11.14. Schematic plot of free energies for α, β, and metastable β' phases. Note that the solubility of B in α is greater for the metastable β'.

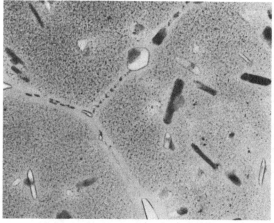

11.15. Precipitation-free zones at grain boundaries in an aluminum-base alloy. From D. A. Porter and K. E. Easterling, *Phase Transformations in Metals and Alloys*, 2nd ed. (London: Chapman & Hall, 1992), p. 306.

precipitate particles gradually decreases as the larger particles grow and the smaller ones shrink and disappear. The driving force for this coarsening is the decrease of surface area between the precipitate and the matrix for large particles. The surface energy per volume is lower for large precipitate particles than for small ones so the amount of solute in solution near large particles is less than that near small ones. This results in a concentration gradient that allows solute to diffuse from the small particles to the large ones, as illustrated in Figure 11.17. This coarsening process is called *Ostwald ripening*.

Martensitic transformations

Martensitic transformations occur by nucleation and shear. There is no composition change and hence no growth by diffusion. Rather, a region of the lattice

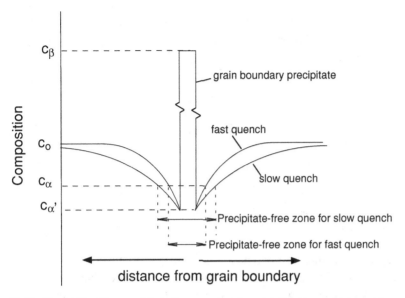

11.16. The width of the precipitate-free zone can be minimized by rapid quenching. $c_{\alpha'}$ and c_{α} are the equilibrium compositions at high and low temperatures, respectively.

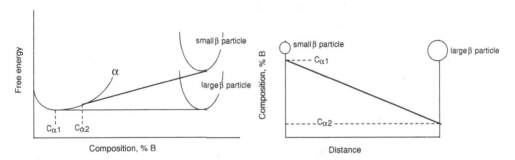

11.17. The higher free energy of small particles (upper left) results in a greater solubility of B in α. This in turn produces a concentration gradient in α between small and large β particles (right), which causes the diffusion needed for the small particles to dissolve and to allow the large ones to grow.

suddenly transforms by shear. The elastic energy caused by misfit of the new and old lattice is minimized if the region undergoing the transformation is lenticular. Figure 11.18 shows the effect of a shear strain of $\gamma = 2$ on a spherical particle (top) and an ellipsoidal particle of the same volume (bottom). With the ellipsoidal particle, there is much less disturbance of the surrounding matrix.

A martensitic transformation occurs over a temperature range. The temperature at which the martensite first starts to form on cooling is called the M_s temperature. More martensite will form only if the temperature is lowered. The temperature at which the reaction is complete is called the M_f temperature. However, the concept of an M_f temperature may be more of a convenience than a reality because often there is no sharp completion of martensite formation.

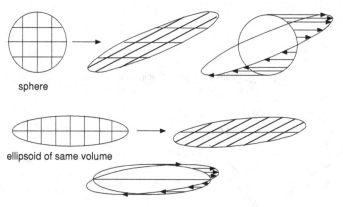

11.18. A spherical region undergoing a martensitic shear distorts a much larger volume of material than an ellipsoid of the same volume.

In most systems the martensitic reaction is geometrically reversible. On heating, the martensite will start to form the higher temperature phase at the A_s temperature and the reaction will be complete at an A_f temperature, as illustrated in Figure 11.19. Martensite in the iron–carbon system is an exception. On heating, the iron–carbon martensite decomposes into iron carbide and ferrite before the A_s temperature is reached. Martensite can be induced to form at temperatures somewhat above the M_s by deformation. The highest temperature at which this can occur is called the M_d temperature. Likewise, the reverse transformation can be induced by deformation at the A_d temperature somewhat below the A_s. The temperature at which the two phases are thermodynamically in equilibrium must lie between the A_d and M_d temperatures.

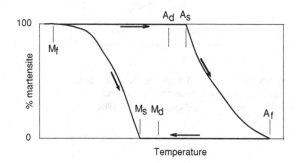

11.19. The fraction martensite increases as the temperature is lowered below the M_s to the M_f. On heating, the reversion starts at A_s and finishes at A_f.

The volumes of the low and high temperature phases are usually not the same so there is often a volume change associated with the martensite reaction.

Spinodal decomposition

Spinodal reactions may occur in systems that have a miscibility gap. On cooling, a single solid solution decomposes into two solid solutions, as indicated in Figure 11.20. The corresponding free energy versus composition curves are shown in Figure 11.21. Consider Alloy 1, cooled suddenly from the single-phase region

PHASE TRANSFORMATIONS

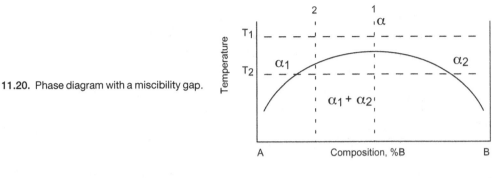

11.20. Phase diagram with a miscibility gap.

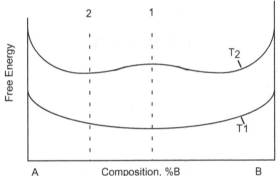

11.21. Free energy versus composition curves for the system in Figure 11.20 at two temperatures.

(temperature T_1) to temperature T_2. Segregation will occur with some regions richer in B and other regions richer in A. This will lower the free energy of the system even though no discrete phase boundary is initially necessary. Because no new boundaries are necessary, there is no nucleation stage in the usual sense of a thermally activated process. The single solid solution can start to break up into two phases without nucleation. Diffusion is, however, necessary.

Figure 11.22 shows that for compositions for which $\partial^2 G/\partial c^2 > 0$, such spinodal decomposition is not possible and nucleation of α_1 and α_2 is required. Hence, spinodal reactions are possible only for composition/temperature combinations for which $\partial^2 G/\partial c^2$ is negative. Sometimes this region is shown by dotted lines on phase diagrams, as in Figure 11.23.

11.22. If Alloy 1, for which $\partial^2 G/\partial c^2 < 0$, segregates into two regions, one rich in B and the other lean in B, the free energy of the system is lowered. On the other hand, if Alloy 2, for which $\partial^2 G/\partial c^2 > 0$, segregates, the free energy of the system increases.

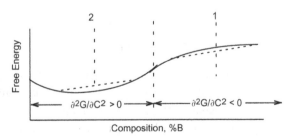

The local composition changes during spinodal decomposition and precipitation by nucleation and growth are compared in Figure 11.24. It is interesting to note that spinodal decomposition requires uphill diffusion. The boundary between

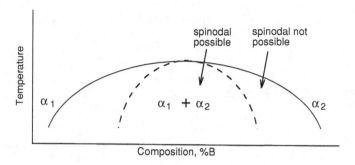

11.23. Phase diagram corresponding to Figure 11.21, showing the region in which a spinodal reaction is possible at temperature T_1.

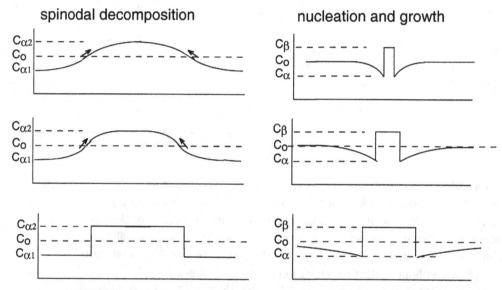

11.24. Comparison of spinodal decomposition and precipitation by nucleation and growth.

the two phases sharpens during the spinodal reaction. In contrast, for nucleation and growth precipitation, the boundaries between the two phases are already sharp.

NOTE OF INTEREST

At room temperature tin has a body-centered tetragonal structure. This is called β or *white tin*. Below 13 °C, the equilibrium structure is diamond cubic and is called *gray tin*. The transformation of white tin to gray tin results in disintegration into a powder because the volume expansion is very large (27%) and the gray-tin phase is very brittle. It would ruin any part made from or joined by tin. However, the transformation is extremely sluggish and inhibited by common impurities. There is an apocryphal story that attributes Napoleon's defeat at Moscow to this transformation. It is said that the cold Russian winter caused the tin buttons on the French uniforms to transform and disintegrate so the French could not fight and hold up their trousers at the same time.

REFERENCES

W. F. Hosford. *Physical Metallurgy.* Boca Raton, FL: CRC Press, 2005.
R. F. Mehl and W. C. Hagel. In *Progress in Metal Physics*, vol. 6. New York: Pergamon, 1956, p. 113.
D. A. Porter and K. E. Easterling. *Phase Transformations in Metals and Alloys*, 2nd ed. London: Chapman & Hall. 1992.
P. G. Shewman. *Transformations in Metals.* New York: McGraw-Hill. 1969.

PROBLEMS

1. The ratio of the activation energy for heterogeneous nucleation of a new phase on a grain boundary, ΔG^*_{hetero}, to that for homogeneous nucleation, ΔG^*_{homo}, is given by $S(\theta) = \Delta G^*_{hetero}/\Delta G^*_{homo} = (1/2)(2 + \cos\theta)(1 - \cos\theta)^2$, where θ is the wetting angle between the new phase, β, and the old one, α.
   A. Calculate $S(\theta)$ if $\gamma_{\alpha\beta} = \gamma_{\alpha gb}$.
   B. According to Figure 11.2, what would be the value of $S(\theta)$ for nucleation at the corners of grains if $\gamma_{\alpha\beta} = 1.2\gamma_{\alpha gb}$?

2. Figure 11.25 gives data on a phase transformation.
   A. Determine the exponent in the Avrami equation.
   B. At what time would you expect the fraction transformed to be 0.001?
   C. At what time would you expect the fraction transformed to be 0.999?

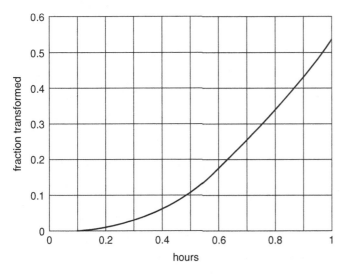

11.25. A plot for a phase transformation of the fraction transformed as a function of time.

3. For a material that undergoes a martensitic transformation, which phase, martensite or austenite, has the lower free energy:
   A. a temperature between the M_s and M_d?
   B. a temperature between the A_s and A_d?

4. A. Describe the conditions under which a reaction $\alpha \rightarrow \alpha_1 + \alpha_2$ will take place by a spinodal decomposition.
   B. Describe the temperature dependence of these conditions.

5. Data from a near eutectoid steel indicate that the growth rate of pearlite is 10^{-2} mm/s at both 660 °C and 540 °C. Using this information, determine the activation energy for the effective diffusion. Take T_e as 723 °C for this steel.

12 Surfaces

Every surface has an energy per area associated with it because molecules or atoms at a surface have different surroundings than those in the interior. The units of surface energy are J/m². Surface tension, in N·m, is equivalent to surface energy. The work to create a surface can be thought of as a surface tension on a line, working through a distance.

Relation of surface energy to bonding

An approximate calculation of surface energy can be made by envisioning a surface being formed by mechanical forces across it and calculating the work required to separate the two halves of a crystal (see Figure 12.1). Here the surface energy, γ_s, is given by

$$\gamma_s = (1/2)(\text{work/area}) = (1/2)\int \sigma \, ds, \qquad (12.1)$$

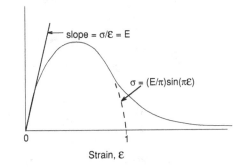

12.1. Schematic plot of bonding strength as a function of atom separation.

where σ is the stress required to create a separation, s, of the two surfaces and the lower and upper limits of integration are $s = 0$ and $s = \infty$. To proceed further along these lines, something must be assumed about how σ varies with s. One assumption is that the curve can be approximated by

$$\sigma = (E/\pi)\sin(\pi \varepsilon), \qquad (12.2)$$

between $\varepsilon = 0$ and 1, where $\varepsilon = (s/s_o)$. Substituting Equation (12.2) and $ds = s_o d\varepsilon$ into Equation (12.1), $\gamma_s = E/(2\pi) \int \sin(\pi\varepsilon) s_o d\varepsilon$. Integrating between $\varepsilon = 0$ and 1,

$$\gamma_s = E s_o/(2\pi^2). \tag{12.3}$$

It is customary to take $s_o = d$ (the atomic diameter). Evaluating this approximation for copper ($E = 128$ GPa and $d = 0.255$ nm), $\gamma_s = 1.65$ J/m^2, compared to the experimentally measured value of 1.7 J/m^2.

Another way of estimating surface energy is based on a hypothetical experiment in which two new surfaces are formed by breaking of bonds. The energy of the two surfaces equals the energy expended in breaking the bonds. One way of estimating this is to assume that the value of the latent heat of sublimation (solid → gas transformation), ΔH_s, expressed on a per-atom basis is the energy to break all of an atom's bonds. Then the surface energy can be found by calculating how many bonds per area must be broken to create the surface. Such calculations predict an orientation dependence of surface energy with more densely packed surfaces having the lower energies. This prediction is in accord with experimental observations of the prevalence of low index surfaces when surface tension controls the surfaces present.

Because both surface energies and melting temperatures are related to bonding strength, the surface energy is closely related to melting point, as shown in Figure 12.2.

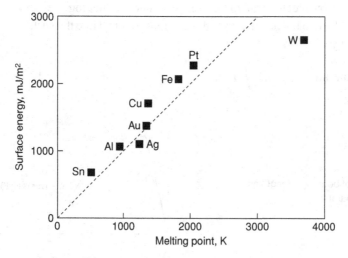

12.2. Correlation of surface energy with melting point. From W. F. Hosford, *Physical Metallurgy*. (Boca Raton, FL: CRC Press, 2004), p. 77, figure 4.8.

Orientation-dependence of surface energy

The surroundings of atoms at the surface of a material are different from those in the interior. A simple way of estimating the orientation dependence of surface energy is to determine the number of missing near-neighbor bonds per area of

SURFACES

surface. The heat of vaporization per atom, h_v, is the energy to break all of an atom's bonds. Therefore, the energy to form a surface is the product of h_v and the number of atoms per area of surface, n_a, and the fraction, α, of their bonds that are missing:

$$\gamma = \alpha h_v n_a. \tag{12.4}$$

Consider a {100} surface of an fcc metal. There are two surface atoms per unit cell so $n = 2/a^2$, and one third of the bonds of each are broken ($\alpha = 1/3$). Breaking these bonds creates two surfaces each of area, a^2. The energy associated with this is $\gamma = 2(1/3)h_v/(2a^2)$. For copper, the heat of sublimation of copper is $(4.73 \text{ MJ/kg})(63.54 \times 10^{-3} \text{ kg/mol})/(6.023 \times 10^{23} \text{ atoms/mol}) = 4.99 \times 10^{-19}$ J/atom, and the lattice parameter is 0.3615×10^{-9} m. Using this data, an estimate of the surface energy is $2(1/3)(4.99 \times 10^{-19} \text{ J/atom})/[2(0.3615 \times 10^{-9})^2\text{m}^2] = 1.27$ J/m². This is lower than the value for polycrystalline copper, 1.7 J/m², but that is expected because the {100} surface energy is lower than the average for a polycrystal.

The surface energy of a crystal is anisotropic. For planes with low indices, it is possible to calculate by trigonometry the number of near-neighbor bonds missing per area and follow the procedure above to estimate the relative surface energies of different low-index planes. The orientation dependence of the free surface energy of a two-dimensional square lattice depends on the orientation of the surface and can be found as follows.

Consider the square lattice in Figure 12.3 and let the area of the surface equal 1. For $0 \leq \theta \leq 90°$, the number of missing vertical bonds is $\sin\theta/a$ and the number of missing horizontal bonds is $\cos\theta/a$, so the total number of missing bonds is $(\sin\theta + \cos\theta)/a$. Breaking these bonds creates two surfaces of total length = 2, so the surface energy per area is

$$\gamma_\theta = U_b(\cos\theta + |\sin\theta|)/(2a), \tag{12.5}$$

where U_b is the energy per bond. This plots as a circle of radius $\sqrt{2}U_b/a$, centered at $x = y = U_b/\sqrt{2}$. Symmetry indicates that in other quadrants γ_θ also plots as a circle of radius $\sqrt{2}U_b/a$. Figure 12.4 is a polar plot of γ.

Gibbs realized that the equilibrium shape of solid is one that minimizes the total surface energy, $U_s = \int \gamma \, dA$. Wulff showed that the minimum energy configuration can be found by constructing planes (Wulff planes) perpendicular to

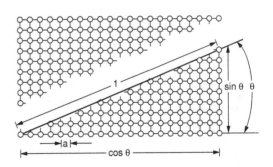

12.3. Two surfaces created by breaking bonds. The surface energy is proportional to the number of broken bonds per area. From W. F. Hosford, *Physical Metallurgy*, (Boca Raton, FL: CRC Press, 2004), p. 79, figure 4.9.

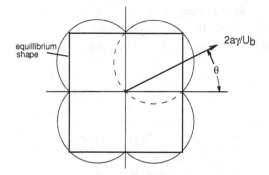

12.4. Polar plot of the surface energy of a two-dimensional square crystal as a function of orientation. The total surface energy of a crystal is a minimum when it is bounded by planes constructed perpendicular to the shortest normals. From W. F. Hosford, *Physical Metallurgy* (Boca Raton, FL: CRC Press, 2004), p. 80, figure 4.10.

radii where γ is a minimum on a polar plot. The equilibrium shape of isolated particles is bounded by portions of Wulff planes that can be reached from the origin without crossing any other Wulff planes. For the two-dimensional crystal in Figure 12.4, this corresponds to a square.

Figure 12.5 is a schematic plot showing how the surface energy changes with θ for orientations near a low-index plane. Plots of surface energy versus orientation have cusps at low-index orientations.

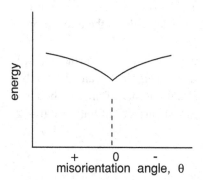

12.5. The dependence of surface energy on the angle of deviation from a low index plane.

A two-dimensional Wulff plot for a three-dimensional crystal having both $\{100\}$ and $\{111\}$ faces is illustrated in Figure 12.6. The corresponding solid is sketched.

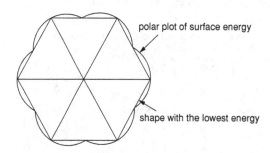

12.6. A polar plot of γ for a hypothetical crystal. The lowest energy shape for the crystal corresponds to planes normal to radii at the cusps.

Surfaces of amorphous materials

A simple bond-breaking approach is not applicable to amorphous materials such as thermoplastics and glasses because they will adjust their molecular configuration to minimize the number of missing bonds. If there is more than one type of bond, the missing bonds at the surface are likely to be the weakest bonds and not characteristic of the overall bond strength.

Grain boundaries

The energy of a grain boundary depends on the misorientation across the boundary. Low-angle tilt grain boundaries are composed of edge dislocations (Figure 12.7). The angle of misorientation, θ, is given by

$$\theta = b/L, \tag{12.6}$$

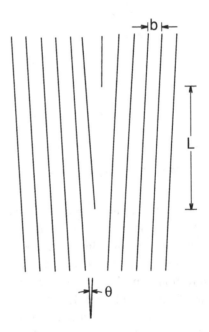

12.7. A low-angle tilt boundary is composed of edge dislocations.

where b is the Burgers vector and L is the distance between dislocations. The number of dislocations per length, n, is equal to the reciprocal of L:

$$n = 1/L = \theta/b. \tag{12.7}$$

The energy of an edge dislocation is given by $(Gb^2/4\pi)\ln(r_1/r_o)/(1+\nu)$, where G is the shear modulus, ν is Poisson's ratio, and r_o is a constant equal to about $b/4$. The value of r_1 is approximately equal to the distance between dislocations, so $r_1 = b/\theta$. Substituting, the energy of a low angle tilt boundary is

$$\gamma = n(Gb^2/4\pi)\ln(r_1/r_o)/(1+\nu) \text{ or } \gamma = (\theta/b)(Gb^2/4\pi)\ln[(4/\theta)]/(1+\nu). \tag{12.8}$$

At low angles, γ is proportional to θ, but at higher angles, γ/θ decreases, as illustrated in Figure 12.8. Screw dislocations on a plane form twist boundaries. The misorientation across a low-angle twist boundary and its energy are proportional to the number of dislocations.

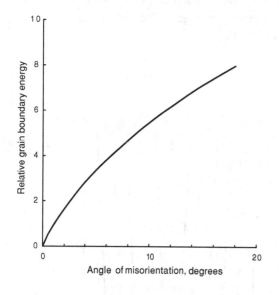

12.8. Relative energy of a low-angle tilt boundary calculated from Equation 12.4.

Certain high-angle boundaries have low energies. For special rotations about <100> and about <111> normals, there is a coincidence of atom sites. Figure 12.9 is a plan view of a 36.9° twist boundary in a simple cubic lattice. One fifth of the atoms have common sites. In an fcc lattice, a 60° rotation about a <111> direction forms a twin boundary. All of the atoms in the boundary have sites that are common to both sides of the twin boundary. The energies of these special boundaries are much lower than the energy of a general high-angle boundary. Experimental measurements of the energies of high-angle boundaries indicate that the energy is between one third and one half of the free surface energy.

Figure 12.10 shows that five independent parameters are needed to describe fully a grain boundary. Three angles are required to characterize the misorientation and two angles are needed to identify the orientation of the grain boundary plane. Studies of grain boundaries in polycrystalline materials have shown that boundaries with the lowest energies have the highest occurrence.* These correspond to

---

* D. M. Saylor, A. Morawiec, and G. S. Rohrer, *J. Amer. Ceram. Soc.* 85 (2002): 3081–3 and D. M. Saylor, A. Morawiec, and G. S. Rohrer, *Acta Mater.* 51 (2003): 3363–86.

SURFACES

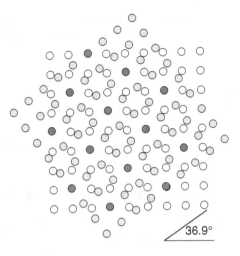

12.9. Coincidence grain boundary in a simple cubic lattice that corresponds to a 36.9° rotation about a <100> direction. One fifth of the atoms (dark circles) in the boundary have common sites.

certain boundary planes. The most frequent boundary planes in MgO are near {100}. In fcc metals the most frequent boundary planes are near {111}, but a number also correspond to {110} tilt boundaries. Coincidence boundaries were not frequently observed.

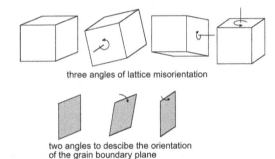

three angles of lattice misorientation

12.10. Five angles are required to describe a grain boundary.

two angles to descibe the orientation of the grain boundary plane

Segregation to surfaces

Segregation of solutes to a grain boundary lowers its energy. Solutes that have a larger atomic size than the parent atoms occupy positions that are open and smaller solute atoms occupy sites where there is crowding, as shown schematically in Figure 12.11.

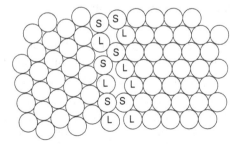

12.11. Two-dimensional sketch of a grain boundary. The crowding at atom positions indicated by S can be relieved if they are occupied by small atoms. Open positions, indicated by L, attract large atoms.

The ratio of grain boundary concentration to overall concentration, $c_gb/c_{\phi 0}$, decreases with increasing solubility. Hondras and Seah* showed that

$$c_gb/c_0 = A/c_{\max}, \tag{12.9}$$

where $c_{\max}$ is the solubility limit in the matrix and A is a constant with a value of about 1.

There is also segregation of solutes to free surfaces. One example is soapy water. Soap segregates to the surface, lowering the surface energy (surface tension).

Direct measurements of surface energy

Most measurements of surface energies have been indirect, comparing the energy of one surface to that of another. There have been relatively few direct measurements of surface energy. A classic experiment in some physics courses is the measurement of the surface tension of soapy water. This is illustrated in Figure 12.12. Measurement is made of the force, F, that must be applied to a soap film of fixed length, L, to keep it from contracting. A force balance gives $F = 2L\gamma$, so the surface tension is given by

$$\gamma = F/(2L). \tag{12.10}$$

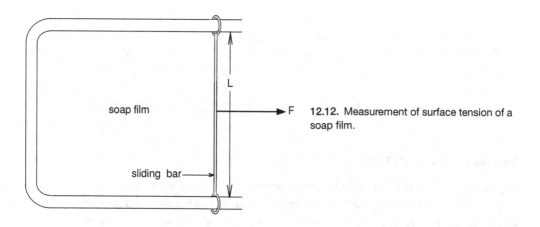

12.12. Measurement of surface tension of a soap film.

The reason that F is divided by $2L$ is that there are two liquid–vapor interfaces associated with a soap film.

Buttner, Udin, and Wulff† measured the surface energy for gold (or more properly, the energy of the interface between solid gold and its vapor). A weighted gold wire was suspended from the cover of an evacuated gold box (Figure 12.13), the system was heated to a high temperature, and the rate of creep was measured.

---

* E. D. Hondras and M. P. Seah, *Int. Metall. Rev.* 42 (1977): Review No. 222.
† F. H. Buttner, H. Udin, and J. Wulff, *J. Metals* 3 (1951): 1209.

SURFACES

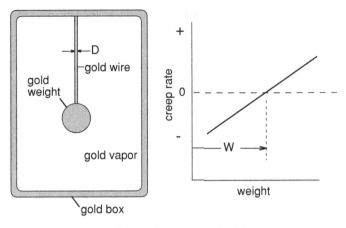

12.13. Measurement of the surface energy of gold.

The experiment was repeated with different weights. With low weights, the wire contracted because of its surface tension, but with greater weights, the wire elongated. They determined the weight, W, at which the wire neither elongated nor contracted. One might be tempted to calculate the surface tension, γ, by equating the weight, W to the longitudinal force, $\pi D \gamma$, where D is the wire diameter and πD is its circumference. However, this would neglect the fact that the surface tension is also acting to contract the circumference. The surface area of the wire is $A = \pi DL$. As the wire elongates by dL, its diameter changes by dD so the area changes by $dA = \pi D dL + \pi L dD$. Because the volume, $V = \pi D^2 L/4$, is constant, $dV = (\pi/4)(D^2 dL + 2LD dD) = 0$, or $dD = -(D/2L)dL$. Substituting this in the expression for the area, $dA = \pi[DdL - (D/2)dL]$, so $dA = (\pi D/2)dL$. The change of total surface energy is γdA and this must equal the external work, $W dL$. Therefore, $\gamma(\pi D/2)dL = W dL$, or

$$\gamma = 2W/\pi D. \tag{12.11}$$

Values of surface energy measured by this technique were found to be 1.140 J/m² for Ag at 903 °C, 1400 J/m² for Au at 1204 °C, and 1.650 J/m² for Cu at 1000 °C. The solid–vapor surface energy is relatively independent of the temperature, but the surface energy of a solid–vapor interface depends on its crystallographic orientation, so these measured values must reflect an average value for many orientations.

Measurements of relative surface energies

Most surface energies have been determined from other surface energies by measuring the angles at which surfaces meet. The angles at which three surfaces meet depend on the relative energies of the three interfaces (Figure 12.14). Each surface exerts a force per length on the junction that is equal to its surface tension. At equilibrium, the force vectors must form a triangle, so the law of sines gives

$$\gamma_{23}/\sin\theta_1 = \gamma_{31}/\sin\theta_2 = \gamma_{12}/\sin\theta_3. \tag{12.12}$$

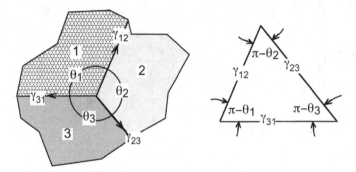

12.14. Relative surface energies.

Often two of the angles and two of the surface energies are equal. For example, consider the intersection of a grain boundary with a free surface. If the temperature is high enough and the time is long enough, the surface will thermally etch (by vaporization or surface diffusion) until an equilibrium angle is formed. From a balance of forces parallel to the grain boundary (Figure 12.15),

$$\gamma_{gb} = 2\gamma_{sv}\cos(\theta/2). \tag{12.13}$$

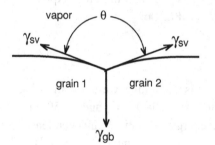

12.15. Intersection of a grain boundary with a free surface.

Similar relations can be used to find the twin boundary energy from the grain boundary energy and the boundary energy between two phases from the energy of a grain boundary in one of them.

Wetting of grain boundaries

Grain boundaries of one phase, α, will be completely wetted by a second phase, β, if $\gamma_{gb} \geq 2\gamma_{\alpha\beta}$. In this case $\theta = 0$. If $0 < 2\theta < 60°$, the edges and corners of

SURFACES

Table 12.1. Surface energies

	γ_{SV} (J/m²)	γ_{gb} (J/m²)	γ_{LS} (J/m²)	γ_{gb}/γ_{SV}	γ_{gb}/γ_{SV}
Copper	1.70	0.60	0.181	0.35	0.106
Silver	1.40		0.135		0.096
Gold	1.20	0.40	0.128	0.33	0.107

Source: From J. H. Brophy, R. M. Rose, and J. Wulff, *Thermodynamics of Structure* (New York: Wiley, 1964), p. 1870. Reprinted with permission of John Wiley & Sons, Inc.

grains will be wetted by β if $0 < 2\theta < 60°$ and the β phase will form a continuous network along the α grain boundaries. If $60 < 2\theta < 120°$, only the grain corners of α will be wetted by β.

Relative magnitudes of energies

Table 12.1 gives the surface energies of the fcc metals copper, gold, and silver. Since about one third of the near-neighbor bonds are missing at a free surface, it can be concluded that at a grain boundary about 10% of the near-neighbor bonds are missing and at a liquid–solid interface roughly 3.5% of the bonds are missing.

NOTE OF INTEREST

As children, most of us were fascinated by soap bubbles. One cannot blow bubbles of pure water. We were told that soap made bubbles possible because soap lowered the surface tension by segregating to the surface of the water. However, it is not the surface tension, per se, that permits bubbles to be blown from soapy water. It is because the surface tension of soap solutions is variable. The surface tension at the top of a bubble is higher than that at the bottom because of the weight of the water it must support. This requirement can be met in soap solutions by different surface concentrations at different locations.

REFERENCES

J. H. Brophy, Robert M. Rose, and J. Wulff. *Thermodynamics of Structure*. New York: Wiley, 1964.
W. F. Hosford. *Physical Metallurgy*. Boca Raton, FL: CRC Press, 2005.
D. A. Porter and K. E. Easterling. *Phase Transformations in Metals and Alloys*, 2nd ed. London: Chapman & Hall, 1992.

PROBLEMS

1. Estimate for a simple cubic crystal the relative energies of a {410} and {420} face. (Find $\gamma_{410}/\gamma_{420}$.)
2. A. Consider a {111} face of an fcc crystal. For the surface atoms, what fraction of the near-neighbor bonds is missing?

B. Consider a {100} face of an fcc crystal. For the surface atoms, what fraction of the near-neighbor bonds is missing?

C. A reasonable first-order approximation is that the surface energy is proportional to the number of missing bonds per area. Using this assumption, estimate $\gamma_{111}/\gamma_{100} \cdot \gamma_{111}/\gamma_{100}$.

3. For copper, what is the equilibrium angle between
   A. a free surface and a grain boundary?
   B. a grain boundary and a twin boundary?

   The surface energy of a twin boundary in copper is about 20 mJ/m.^{2}

4. Evaluate $\gamma_\theta/(U_b/a^2)$ for a two-dimensional crystal with a square lattice over the range $-10° \leq \theta \leq +10°$ and plot $\gamma_\theta/(U_b/a^2)$ versus θ.

5. Calculate the pressure inside a sphere of solid gold having a diameter of 20 μm.

6. Estimate the free surface energy of lead and nickel. Lead melts at 327 °C and nickel melts at 1455 °C.

7. Knowing that low-angle tilt boundaries are composed of edge dislocations, predict the most common planes for tilt boundaries in fcc and in bcc metals.

8. Examine the copper–aluminum phase diagram and predict whether c_{gb}/c_0 is larger for copper segregating to grain boundaries in aluminum or aluminum segregating to grain boundaries in copper.

9. The microstructure of superalloys consists of fcc γ' particles in a matrix of fcc γ. The γ' particles are usually cuboidal. Speculate as to the reason for this morphology.

13 Bonding

Ionic binding energy

Ionic bonding is the result of the mutual attraction of ions of opposite charge. The energy of the bond, U_{pair}, between a pair of oppositely charged ions depends on the charges on the ions and their separation,

$$U_{\text{pair}} = -z_1 z_2 e^2 / d, \qquad (13.1)$$

where $z_1 e$ and $z_2 e$ are the charges on the ions and d is their separation. For a crystal, the bonding energy for an ion is the sum of the attractions and repulsions to all of the other ions in the crystal. This can be expressed as

$$U = -M z_1 z_2 e^2 / d, \qquad (13.2)$$

where M is the Madelung constant. The value of M can be calculated by considering the attraction and repulsion between a central ion and each shell of ions about it. For example, the first two shells of ions in sodium chloride about a central positive ion are sketched in Figure 13.1. In the first shell around a central positive ion, there are six negative ions at a distance d, twelve positive ions at a distance

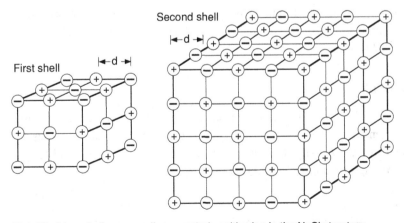

13.1. First two shells surrounding a central positive ion in the NaCl structure.

Table 13.1. Madelung constants

Structure	Madelung constant
Sodium chloride	1.74755
Cesium chloride	1.76267
Zinc blende	1.63806
Wurtzite	1.641

$\sqrt{2}d$, and eight negative ions at a distance $\sqrt{3}d$, so the bonding energy with the first shell is $-(-6/1 + 12/\sqrt{2} - 8/\sqrt{3})z^2e^2 = +2.1335z^2e^2/d$.

For the second shell the bond energy is $-(+6/2 - 24/\sqrt{5} + 24/\sqrt{6} + 48/\sqrt{8} - 48/\sqrt{9} + 8/\sqrt{12})z^2e^2 = -3.1405z^2e^2/d$. Similarly, the bonding energies with the third through seventh shells are $+3.2797, -0.74295, +0.4125, -0.1937932$, and $+0.1658019$ times z^2e^2. Continued calculations for subsequent shells form a series that converges at $M = 1.747$, as illustrated in Figure 13.2. Table 13.1 gives the Madelung constants for several crystal structures.

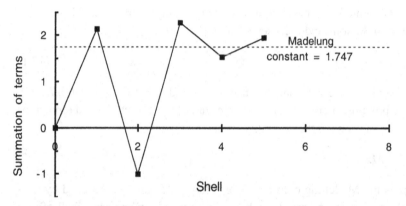

13.2. Sum of the attractive energy of subsequent shells in a NaCl crystal. The series converges at $M = 1.74755$.

Melting points

Since $U = -z_1z_2e^2/d$, the binding energies of ionic crystals of formula AB increase with the square of z. The melting points of ionic solids correlate very well with z^2/d (Figure 13.3).

Elastic moduli

The elastic moduli of ionic crystals are also related to the valences and the distances between ions. The force, $F = dU/dd$, required to separate two ions is proportional to z_1z_2/d^2. The stress is $\sigma = F/A$, where A is proportional to d^2. Therefore, Young's modulus, E, of AB compounds is

$$E = Cz^2/d^4, \qquad (13.3)$$

where C is a constant that reflects the Madelung constant, α. Figure 13.4 shows that the elastic constant for AB compounds is approximately proportional to z^2/d^4.

BONDING

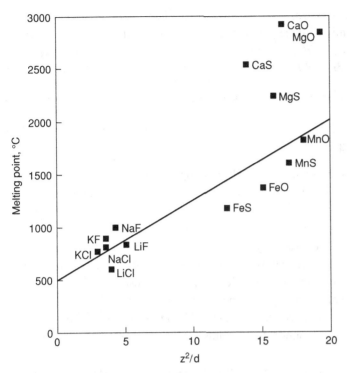

13.3. The melting points of AB ionic crystals increase with z^2/d.

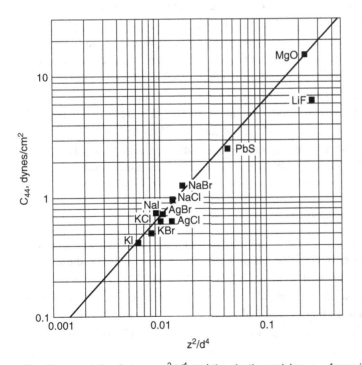

13.4. The correlation between z^2/d^4 and the elastic modulus, c_{44}, for various AB compounds with a NaCl structure. Data from J. J. Gilman, *Progress in Ceramic Science* 1 (1961): 146–94.

The ionic bonding energy is $U = -\alpha z^2 e^2/d$. The bonding force is $F = dU/dd$. In the bonding stress, $\sigma = F/A$, A is proportional to d^2 so the bonding stress $\sigma = \beta(1/d^2)\partial u/\partial r$, where $\beta = d^2/A$. Substituting, $d\sigma/dr = (1/d^2)\partial^2 u/\partial d^2$ and $d\varepsilon = dd/d$,

$$E = d\sigma/d\varepsilon = (1/d)(\partial^2 u/\partial d^2) = CZ^2/d^4. \tag{13.4}$$

Covalent bonding

Bond strengths, elastic moduli, and melting temperatures for covalent bonding also all increase with z^2 and decrease with d. Silicon carbide, silicon nitride, and aluminum oxide all have a very high stiffness and melting point. There are also repulsive forces between ions caused by overlapping of the inner electron clouds. These are appreciable only over very short distances. The usual approximation is that the repulsive energy between two ions is proportional to $1/D^n$, where n is approximately 9, so this term is negligible except for nearest neighbors.

Covalent bonds involve sharing of valence electrons to complete electron shells. For each bond, two electrons are shared. Covalent bonds are generally tighter than ionic bonds. In many compounds the bonding is intermediate between ionic and covalent.

Hydrogen atoms have one valence electron. Covalently bonded to another atom (e.g., oxygen or carbon), it shares its electron and one of the other atom's to form a bond. Carbon has four valence electrons. When bonded in CH_4 or C_2H_6, it shares two electrons with each of the hydrogen atoms and the other carbons. In ethylene, C_2H_4, the carbon atoms share four electrons between them, forming a double bond. Table 13.2 gives the valences of several elements in organic compounds.

Bonds have characteristic strengths and lengths. Table 13.3 gives the energies of several bonds and the bond lengths in organic compounds.

When C and Si are bonded to four other atoms of the same specie (e.g., diamond, CH_4, CCl_4, silicon, SiF_4), the bonds are at tetrahedral angles of $\arccos(1/3) = 109.5°$. When all of the surroundings are not the same (e.g., CH_3Cl, C_2H_6), angles deviate from $109.5°$. Table 13.4 gives the bond angles for some bonds.

Geometric considerations

The structure of ionic crystals usually corresponds to the maximum possible coordination number. If the ions are assumed to be hard spheres, there must be contact between ions of opposite signs and no contact between ions of like sign. The coordination depends on the ratio of anion-to-cation diameters. Figure 13.5 shows that the critical condition for threefold coordination is $R/(R+r) = \cos 30° = \sqrt{3}/2$. Therefore, to prevent contact between like ions,

$$r/R > 0.1547. \tag{13.5}$$

BONDING

Table 13.2. Valences

Element	Valence
H, Cl, F, B	1
O, S, Se, Te	2
N, P, As	3
C, Si, Ge	4

Table 13.3. Energies and bond lengths

	Bond energy (kJ/mole)	Bond length, nm
C–H	435	0.11
C–C	370	0.155
C≡C	680	0.13
C=C	890	0.12
O–H	500	0.10
C–O	360	0.14
C=O	535	0.12
C–F	450	0.14
C–Cl	340	0.18
N–H	430	0.10
O–O	220	0.15
H–N	435	0.074

Table 13.4. Bond angles (degrees)

Diamond	C	C–C–C	109.5
Methane	CH_4	H–C–H	109.5
Ammonia	NH_3	H–N–H	107
Water	H_2O	H–O–H	104
Hydrogen sulfide	H_2S	H–S–H	92

Source: Data from W. G. Moffatt, G. W. Pearsall, and J. Wulff, *Structure* (New York: Wiley, 1964). Reprinted with permission of John Wiley & Sons, Inc.

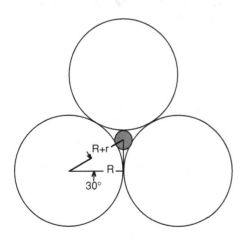

13.5. For threefold coordination, $r/R > 0.155$.

Fourfold or tetrahedral coordination corresponds to the smaller ion at the center of a tetrahedron with the larger ions on the corners, as shown in Figure 13.6. The critical condition can be easily analyzed by imagining the tetrahedron inside a cube. Then $[2(R + r)]^2 = 3a^2$. At the critical condition, $(2R)^2 = 2a^2$, so $4(R + r)^2 = 6R^2$, and

$$r/R > \sqrt{3/2} - 1 = 0.2247. \tag{13.6}$$

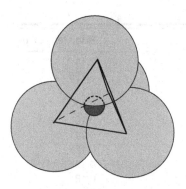
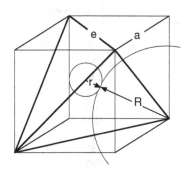

13.6. For fourfold tetrahedral coordination, $r/R > 0.2247$.

Figure 13.7 shows that sixfold or octahedral coordination corresponds to the smaller ion at the center of an octahedron with the larger ions on the corners. The critical ratio corresponds to $[2(r + R)]^2 = 2R^2$

$[2(r + R)]^2 = 2(2R)^2$, so

$$r/R > \sqrt{2} - 1 = 0.4142. \tag{13.7}$$

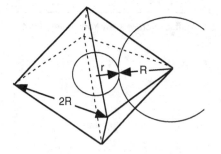

13.7. For sixfold (octahedral) coordination, $r/R > 0.4142$.

With eightfold or cubic coordination, the smaller ion is in the center of a cube with larger ions on the corners (Figure 13.8). $[2(r + R)]^2 = 3(2R)^2$, so

$$r/R > \sqrt{3} - 1 = 0.7321. \tag{13.8}$$

These geometric restrictions are summarized in Table 13.5.

BONDING

Table 13.5. Geometric restrictions on coordination

Coordination	minimum r/R ratio
3-fold	0.1547
4-fold	0.2247
6-fold	0.4142
8-fold	0.7321

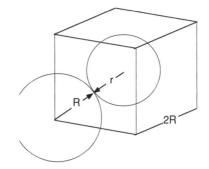

13.8. For eightfold (cubic) coordination, $r/R > 0.7321$.

Ionic radii

When anions are formed, atoms lose electrons. Therefore, the ionic radii of elements are smaller than the corresponding atomic radii. Both ionic and atomic radii of elements in the same period decrease with increasing atomic number. See Figure 13.9.

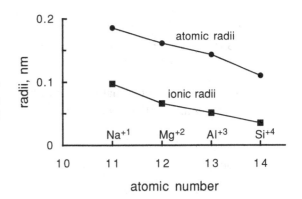

13.9. For metals, the ionic radii are smaller than the corresponding atomic radii. For the same period, both decrease with increasing atomic number.

For the same column of the table, both the anion and atomic radii increase with the atomic number, as shown in Figure 13.10. Cation radii are larger than the corresponding atomic radii and, for the same column of the periodic table, both the cation and atomic radii increase with the atomic number, as shown in Figure 13.11.

Generally anions have smaller radii than cations. Ionic radii of anions tend to decrease with higher valences. Table 13.6 lists ionic radii for sixfold coordination.

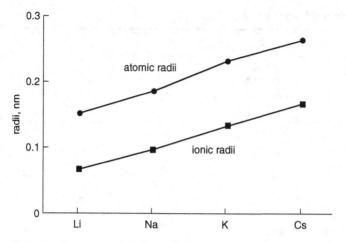

13.10. For the same period, the radii of metal atoms and anions increase with atomic number.

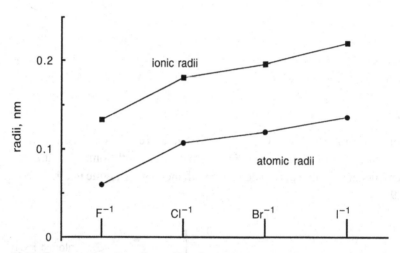

13.11. For nonmetals, the cation radii are larger than the atomic radii. For the same period, both increase with atomic number.

Ionic radii are somewhat smaller with fourfold coordination and are somewhat larger with eightfold coordination:

$$R_{CN=4} = R_{CN=6}. \qquad (13.9)$$

and

$$0.97 R_{CN=8} = R_{CN=6}. \qquad (13.10)$$

Structures of compounds

Many ionic compounds have structures based on either fcc or hcp packing of one of the ions. Both of these structures have sites of fourfold coordination and sites of sixfold coordination. Several simple structures for AB compounds are illustrated in Figure 13.12. Both the zinc blende structure, which is based on

BONDING

Table 13.6. Ionic radii

Ion	Radius, nm	Ion	Radius, nm
Li +1	0.068	Be +2	0.035
O −2	0.140	F −1	0.133
Na +1	0.097	Mg +2	0.066
Al +3	0.051	Si +4	0.042
S −2	0.184	Cl −1	0.181
K +1	0.133	Ca +2	0.099
Ti +4	0.068	Cr +3	0.063
Mn +2	0.074	Fe +2	0.074
Fe +3	0.064	Co +2	0.072
Ni +1	0.069	Cu +2	0.096
Zn +2	0.074	Ag +1	0.126
Sn +4	0.071	I −1	0.220
Cs +1	0.167	W +4	0.070
Au +1	0.137	Hg +2	0.110
Pb +2	0.120	U +4	0.097

Source: Data from L. Pauling, *Nature of the Chemical Bond*, 3rd ed. (Ithaca, NY: Cornell Univ. Press, 1960), p. 514, table 13.3.

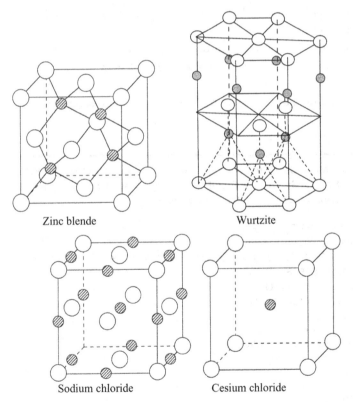

13.12. Several simple structures for AB compounds.

Table 13.7. Common structures

Anion packing	Coordination		Cation sites	Name	Example
	Anion	Cation			
FCC	6	6	All oct	Halite	NaCl
FCC	4	4	1/2 tet	Zinc blende	αZnS
FCC	4	8	All tet	Antifluorite	Li_2O
FCC	12,6	6	1/4 oct	Perovskite	$CoTiO_3$
FCC	4,6	4	1/8 oct	Spinel	$FeAl_2O_4$
HCP	6	6	All oct	Nickel arsenide	NiAs
HCP	4	4	1/2 tet	Wurtzite	βZnS
HCP	6	4	2/3 oct	Corundum	Al_2O_3
HCP	6,6	4	2/3 oct	Ilmenite	$FeTiO_3$
HCP	6,4	4	1/2 oct	Olivine	Mg_2SiO_4
Simple cubic	8	8	All cubic	Cesium chloride	CsCl
Simple cubic	8	4	1/2 cubic	Fluorite	CaF_2

an fcc arrangement of cations, and the wurtzite structure, which is based on an nhcp arrangement of cations, have fourfold coordination. The sodium chloride structure, which is based on an fcc arrangement of cations, has sixfold coordination. The eightfold coordination of the cesium chloride structure is based on a simple cubic arrangement of cations.

The packing of anions, the coordination of the anions and cations, and the cation site occupation for a number of the common crystal structures are listed in Table 13.7

Table 13.8 is a compilation of the structures of many inorganic compounds.

Table 13.8. Structures of some compounds

Halite	NaCl, MgO, CaO, SrO, BaO, CdO, MnO, FeO, CoO, NiO KCl, KI, KBr,
Zinc blende	αZnS, ZnO, SiC, BeO, AlP, GaP, αCdS, HgS, βAgI, InP, BeSe, AlAs, GaAs, CdSe, HgSe, CuI, InSb, BeTe, AlSb
Antifluorite	Li_2O, Na_2O, K_2O
Perovskite	$CoTiO_3$, $SrTiO_3$, $SrSnO_3$, $BaTiO_3$, $SrZrO_3$
Spinel	$FeAl_2O_4$, $ZnAl_2O_4$, MgA_2O_4
Wurtzite	βZnS, ZnO, SiC, BeO, MgTe, CdSe, αAgI, βZnS, βCdS, MgTe
Nickel arsenide	NiAs, FeS, FeSe, CoSe
Corundum	Al_2O_3, Fe_2O_3, Cr_2O_3, Ti_2O_3, V_2O_3
Ilmenite	$FeTiO_3$, $NiTiO_3$, $CoTiO_3$
Olivine	Mg_2SiO_4, Fe_2SiO_4
Cesium chloride	CsCl, CsBr
Fluorite	CaF_2, UO_2, CeO_2, ZrO_2, HfO_2

NOTE OF INTEREST

Linus Pauling (1901–1994) was one of the twentieth century's greatest scientists. He twice won Nobel prizes, once for chemistry and once for peace. His *Nature*

BONDING

of the Chemical Bond and the Structure of Molecules and Crystals (Ithaca, NY: Cornell Univ. Press, 1960) treated bond angles and lengths and has become the basis for predicting coordination in crystals.

REFERENCES

J. J. Gilman. *Progress in Ceramic Science* 1 (1961): 146–99.
W. D. Kingery. *Introduction to Ceramics*. New York: Wiley, 1960.
W. D. Kingery, H. K. Bowen, and D. R. Uhlmann. *Introduction to Ceramics*, 2nd ed. New York: Wiley, 1960.

PROBLEMS

1. Calculate the Madelung constant for a two-dimensional crystal with a square unit cell having a positive ion at the center and negative ions on the corners.
2. A. Calculate the energy released as heat when 1 g of ethylene ($CH_2=CH_2$) is polymerized to polyethylene ($-CH_2-CH_2-$).
   B. If the process were adiabatic, what would be the temperature rise? The heat capacity of polyethylene is 2 kJ/kg °C.
3. Estimate Young's modulus for NaCl given that Young's modulus for MgO is 210 GPa.
4. The boron ion, B^{+3}, has a radius of about 0.025 nm. Predict the coordination of B^{+3} in B_2O_3.
5. Predict the structure of CsI.
6. Predict the structure of AgI.
7. The structure of diamond is like the zinc blende structure except that all of the atoms are carbon. Calculate the lattice parameter of diamond if the atomic diameter of carbon is 0.92 nm.
8. What coordination would be expected if $r/R < 0.1547$?

14 Sintering

Sintering is a process of bonding small particles without melting them. It is a simple and cheap way of fabricating parts of metals, ceramics, and some polymers. The driving force for sintering is the reduction energy resulting from decreased surface area. Most ceramics are consolidated by sintering. These include clay products as well as refractory oxides. These ceramics cannot be fabricated by melting and freezing. Sintering is also used to produce parts of metals that are difficult to melt. Examples include carbide tools and tungsten for lamp filaments. Mixed powders are sintered to make composites that are not otherwise possible, such as friction materials for brakes and clutches. Porous parts for filters or oilless bearings are made by incomplete sintering. Even some polymeric materials are sintered. Teflon cannot be melted without decomposing so Teflon parts are made by sintering powder.

Mechanisms

During sintering adjacent particles adhere and a neck is formed at the area of contact. Figure 14.1 is a micrograph of such a neck formed between two nickel spheres. There are two groups of sintering mechanisms, as shown in Figure 14.2. Mechanisms like vapor and surface diffusion transport material from the surface to form the neck. These do not change the distance between the centers of particles so they contribute little to densification. Mechanisms that transport material from the interface between the particles to form the neck (grain boundary and lattice diffusion) do cause densification.

German[*] gives several geometric relations between the particle radius, r; the neck radius, ρ; the movement of the center of the particle toward the plane of contact, h; the area of contact, A; and the volume of material that must be transported to form the neck, V (Figure 14.3).

---

[*] R. M. German, *Sintering Theory and Practice* (New York: Wiley, 1996).

SINTERING

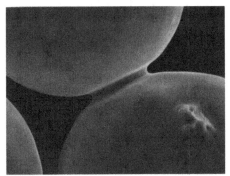

14.1. An SEM photograph of the neck formed between two 33-mm diameter spheres of nickel after 30 min at 1030 °C. The marker is 5 μm. Reprinted with permission from *Powder Metallurgy Science*, 1984, Metal Powder Industries Federation, 105 College Road East, Princeton, New Jersey, USA.

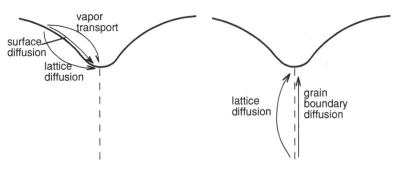

14.2. Growth of a neck by transport from the spherical surface (left) and from the grain boundary formed between the particles (right).

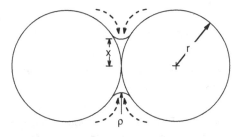

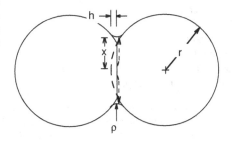

14.3. Two cases of sintering: Transport of material from the spherical surfaces to the neck (top) does not contribute to densification. Transport of material from the interface between the particles to the neck (bottom) does contribute to densification. ρ is the neck's radius of curvature, r is the particle radius, $2h$ is the decrease of distance between particle centers, and x is the radius of contact.

If there is no shrinkage,

$$V = \pi x^4/(2r) \tag{14.1}$$

$$h = 0 \tag{14.2}$$

$$A = \pi^2 x^3/r \tag{14.3}$$

$$\rho = x^2/(2r). \tag{14.4}$$

For the shrinkage case,

$$V = \pi^4/(4r) \tag{14.5}$$

$$h = x^2/(2r) \tag{14.6}$$

$$A = \pi^2 x^3/(2r) \tag{14.7}$$

$$\rho = x^2/(4r). \tag{14.8}$$

Early stage of sintering

In the early stage of sintering, the rate at which the contact increases is described by

$$(x/r)^n = Btr^m \tag{14.9}$$

and the shrinkage is described by

$$(\Delta L/L)^{n/2} = Bt/(2^{n-m} r^m), \tag{14.10}$$

where x and r are defined in Figure 14.3, L is any linear dimension, and the exponents n and m depend on the mechanism. The term B expresses the temperature dependence:

$$B = B_o \exp(-Q/RT). \tag{14.11}$$

Table 14.1 gives the exponents n and m and expressions for B for various mechanisms.

Table 14.1. Exponents for several mechanisms

Mechanism	n	m	B
Viscous flow	2	1	$3\gamma/2\eta$
Plastic deformation	2	1	$9\pi\gamma bD_v/(kT)$
Vapor transport	3	2	$(3P\gamma/\rho^2)(\pi/2)^{1/2}(M/kT)^{3/2}$
Volume diffusion	5	3	$80 D_v \gamma \Omega/(kT)$
Surface diffusion	7	4	$56 D_s \gamma \Omega^{4/3}/(kT)$
Grain boundary diffusion	6	2	$20\delta D_b \gamma \Omega/(kT)$

Note: γ = surface energy, D_v = volume diffusivity, D_s = surface diffusivity, D_b = grain boundary diffusivity, η = viscosity, b = Burgers vector, k = Boltzman's constant, ρ = density, δ = width of grain boundary diffusion path, P = pressure, M = molecular weight, and Ω = atomic volume.
Source: From R. M. German, *Sintering Theory and Practice* (New York: Wiley, 1996). Reprinted with permission of John Wiley & Sons, Inc.

SINTERING

Intermediate stage of sintering

The unfilled edges between the particles form continuous pore paths along the unfilled edges of the particles, as illustrated in Figure 14.4. These paths can be approximated as cylindrical tubes. The porosity, V_p, is

$$V_p = \pi(d_p/G)^2, \tag{14.12}$$

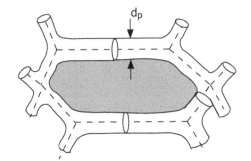

14.4. Continuous tubular pore structure.

where d_p is the diameter of the cylindrical tubes and G is the grain size.

Final stage of sintering

The cylindrical pores become unstable when their length $l \geq \pi d_p$. At this point, they collapse into more or less spherically shaped pores at the grain corners. These spheres have diameters larger than the tubes. The collapse starts at about 15% total porosity and is complete by about 5% total porosity. By assuming that the particles have the shape of tetrakaidecahedra and the pores are spherical, Coble[*] calculated the porosity as

$$V_p = (\pi/\sqrt{2})(d_p/2\,l)^3, \tag{14.13}$$

where l is the length of a side of the tetrakaidecahedron. These pores and those remaining on the edges, faces, and grain interiors are isolated. Therefore, the rate of shrinkage slows. Pores at corners may either shrink or grow, depending on the ratio of surface energies, γ_{SV}/γ_{SS}, where γ_{SV} is the vapor–solid surface energy and γ_{SS} is the solid–solid (grain boundary) surface energy and the number of grains contacted by the pore (Figure 14.5).

Loss of surface area

Throughout the sintering the surface area is decreasing. Figure 14.6 shows the decrease of surface area, $\Delta S/S_o$, of alumina as it is heated.

Once grain growth has allowed the grain boundaries to leave the pores, the only mechanism of densification is by lattice diffusion of atoms from grain boundaries

---

[*] R. L. Coble, *J. Appl. Phys.* 32 (1961): 787–92.

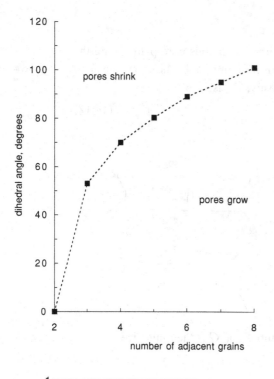

14.5. Whether a corner pore shrinks or grows depends on the dihedral angle and the number of grains it borders. Data from G. C. Kuczynski, in *Powder Metallurgy for High Performance Applications*, ed. J. J. Burke and V. Weiss (Syracuse, NY: Syracuse Univ. Press, 1972).

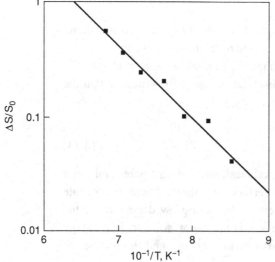

14.6. The loss of surface area as alumina is heated at 5 °C per minute. Data from S. H. Hillman and R. M. German, *J. Mat. Sci.* 27 (1992): 2641–8.

or dislocations to the pores. Lattice diffusion also allows large isolated pores to grow at the expense of smaller pores because the pressure inside the smaller ones is higher.

Particle-size effect

Of course, the time required for sintering is decreased as the particle size decreases. Figure 14.7 shows that if surface diffusion is the controlling mechanism, the time is inversely proportional to D^4.

SINTERING

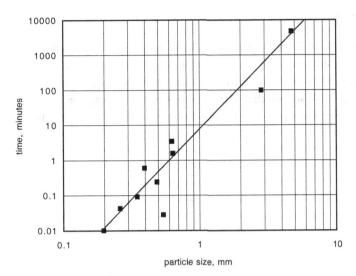

14.7. Time needed to reach a neck size of $x/D = 0.1$ in the sintering of various sizes of ice particles. The controlling mechanism is surface diffusion. Data from W. D. Kingery, *J. Appl. Phys.* 31 (1960): 833–8.

The relation between two temperatures, T_1 and T_2, required for equal degrees of sintering of particles of diameters D_1 and D_2 is

$$1/T_2 = 1/T_1 - m(\text{R}/Q)\ln(D_2/D_1). \tag{14.14}$$

Grain growth occurs during the later stages of sintering. Figure 14.8 shows data for grain growth in alumina at 1550 °C.

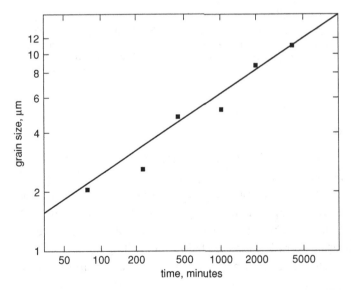

14.8. Grain growth in alumina during sintering at 1550 °C. Data from R. L. Coble, *J. Appl. Phys.* 33 (1961): 793.

Activated sintering

Sometimes the addition of a very small amount of a second material greatly increases the rate of sintering. Usually this can be attributed to the formation of a phase with a much lower melting point in which the diffusion is much faster. Figure 14.9 shows that the sintering rate of tungsten is drastically increased by enough of certain elements to form a four-atom-thick layer.

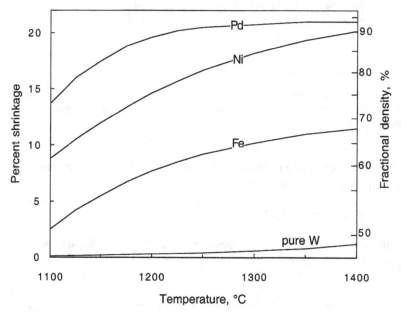

14.9. Sintering of tungsten is activated by addition of small amounts of Pd, Ni, Fe, or other elements. Data are for 1 h at the indicated temperatures. Reprinted with permission from *Powder Metallurgy Science*, 1984, Metal Powder Industries Federation, 105 College Road East, Princeton, New Jersey, USA.

Liquid-phase sintering

Mixtures of powders of two materials sinter very rapidly if one of them melts at the sintering temperature. Initially capillary action causes the liquid phase to rapidly wet the solid phase, causing an initial contraction. Then as the solid phase dissolves in the liquid it is rapidly transported to locations that decrease the pore volume. Carbide tool material is made from a mixture of cobalt and tungsten carbide powders sintered below the melting point of cobalt. The volume fraction liquid must be limited so capillarity can retain the shape during sintering.

SINTERING

Hot isostatic pressing

If pressure is applied at the sintering temperature, plastic deformation by creep increases the rate of sintering. This is referred to by the term HIPping. It is used on castings as well as powder compacts.

NOTE OF INTEREST

Snowballs are the most common example of sintering. Unless it is very cold, ice will sinter rapidly enough to allow snowballs to be made.

REFERENCES

R. M. German. *Powder Metallurgy Science*. Princeton, NJ: Metal Powder Industries Fed., 1984.

R. M. German. *Sintering Theory and Practice*. New York: Wiley, 1996.

G. S. Upadhyaya. *Powder Metallurgy Technology*. Cambridge, U.K.: Cambridge International Science, 1997.

PROBLEMS

1. Stainless steel powder with a mean particle diameter of 50 mm has been compacted to a green density of 58% and sintered in pure H_2. The resulting shrinkage measurements are given below. Published diffusion data for this stainless steel show that the activation energies are 225 kJ/mol for surface diffusion, 200 kJ/mol for grain boundary diffusion, and 290 kJ/mol for volume diffusion. Use the data below to determine the mechanism.

Temperature °C	Time h	Shrinkage %	Temperature °C	Time h	Shrinkage %
1050	2.0	0.62	1200	1.5	1.63
1100	2.0	0.91	1200	2.0	1.82
1150	2.0	1.31	1250	2.0	2.49
1200	0.5	1.05	1300	2.0	3.33
1200	1.0	1.38			

2. A. What is the surface area of 1 g of copper powder with a 50-μm diameter?
   B. What is the total energy of the surface? ($\gamma_{SV} = 1700$ erg/cm^2)

3. The density of a powder after compaction is 85% and after sintering 99%. What diameter die and punch should be used to make a cylinder 25 mm in diameter and 22 mm tall?

4. The time for a given degree of sintering is proportional to D^m. Determine the exponent, m, for ice from the data in Figure 14.7.

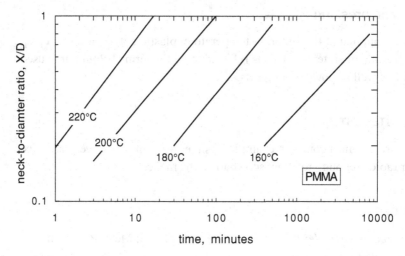

14.10. Neck growth in polymethylmethacrylate at various temperatures. Data from N. Rosenzweig and M. Narkis, *Polymer Sci. Eng.* 21 (1981): 1167–70.

5. Data for the neck growth of polymethyl methacrylate (PMMA) during sintering is plotted in Figure 14.10.
   A. Determine the activation energy.
   B. Determine the value of the exponent, n, in the equation $X/D = Ct^n$.

15 Amorphous Materials

Glass transition

In the amorphous state there is no long-range order and there is no symmetry. There is, however, a great deal of short-range order. If crystallization is prevented during cooling, a glass will form with short-range order inherited from the liquid. The critical cooling rate to prevent crystallization varies greatly from one material to another. See Table 15.1. Glasses form in strongly bonded silicates unless the cooling rate is extremely slow. On the other hand, extremely rapid cooling is required to prevent crystallization of metals. Indeed, glassy structures have been produced only in metallic alloys with complex compositions. These usually have compositions that correspond to deep eutectics.

Table 15.1. Glass transition temperatures and bonding of several glassy solids

Glass	Type of bond	T_g, K
SiO_2	Covalent	1430
As_2Se_3	Covalent	470
Si	Covalent	≈ 800
$Pd_{0.4}Ni_{0.4}P_{0.2}$	Metallic	580
$FeO_{0.82}B_{0.18}$	Metallic	410
$Au_{0.8}Si_{0.2}$	Metallic	290
BeF_2	Ionic	520
Polystyrene	Covalent, van der Waals	370
Se	Covalent, van der Waals	310
Isopentane	Covalent, van der Waals	65
H_2O	Covalent, hydrogen	140
C_2H_5OH	Covalent, hydrogen	90

Source: From R. Zallen, *Physics of Amorphous Solids* (New York: Wiley, 1983), p. 6. Reprinted with permission of John Wiley & Sons, Inc.

The structure of a glass is similar to that of the liquid from which it formed. There is no abrupt change of properties. However, the rate of change of properties with temperature is much less than in the liquid state. For example, consider the volume of a given mass. With crystallization, there is an abrupt volume change, ΔV (usually contraction), as shown in Figure 15.1. On the other hand, if crystallization is prevented, the rate of change of volume with temperature, dV/dT, decreases at the glass transition temperature but there is no abrupt volume change, ΔV.

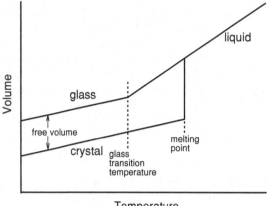

15.1. Schematic representation of the change of volume with temperature for crystallization and glass transition.

The glassy state is usually less dense than the crystalline state. The difference between the volumes is called the free volume:

$$V_{\text{free}} = (V_{\text{glass}} - V_{\text{cryst}})/V_{\text{cryst}}. \tag{15.1}$$

Glass transition in polymers

As the temperature of an amorphous polymer is lowered, there is a transition from rubberlike material with a low Young's modulus to a stiff glass with a high modulus. For example, the Young's modulus of PVC (measured at 1 s) increases from 0.15 to 1.2 GPa as the temperature is decreased from 90 to 75 °C. The glass transition temperature is in this range. The exact temperature depends on the rate of cooling.

Glass transition temperatures of polymers also depend on their structure. They are higher in polymers with inflexible main chain groups and where there are bulky, inflexible side groups. However, long $(CH_2)_n$ side groups lower the glass transition temperature. The addition of plasticizers, which are small molecules, also lowers the glass transition temperature. The effects of di(ethylhexyl)phthalate on the glass transition temperature of PVC are shown in Figure 15.2.

Molecular length

The length of a linear polymer molecule may be described by its contour length, which equals $n\ell$, where n is the number of units of length ℓ. The end-to-end distance, r, of a convoluted polymer molecule is much shorter. It is extremely variable, both with respect to time and to other molecules of the same contour length. To calculate the most probable root-mean-square end-to-end length,

AMORPHOUS MATERIALS

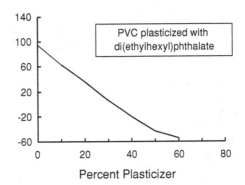

15.2. The glass transition temperature of PVC is lowered by di(ethylhexyl)phthalate.

$<r^2>^{1/2}$, one must assume something about the freedom of rotation at the joints. One extreme assumption is that the molecule is a freely joined chain. In this case

$$<r_f^2>^{1/2} = n^{1/2}\ell. \qquad (15.2)$$

However, there is not complete freedom of rotation at each joint of a linear polymer. Rather, the rotation possible at each joint is governed by the valence angle, θ. With this restriction

$$<r_f^2>^{1/2} = n^{1/2}\ell[(1 - \cos\theta)/(1 + \cos\theta)]^{1/2}. \qquad (15.3)$$

For C–C backbones of polyethylene, $\theta \approx 109.5°$, and $\cos\theta = -1/3$. Substituting into Equation (15.3),

$$<r_{fa}^2>^{1/2} = \sqrt{2} n^{1/2}\ell. \qquad (15.4)$$

If there is steric hindrance from side groups, the mean-square length is

$$<r_o^2>^{1/2} = n^{1/2}\ell\sigma[(1 - \cos\theta)/(1 + \cos\theta)]^{1/2}, \qquad (15.5)$$

where σ is the *steric parameter* that describes the steric hinderance.

The *characteristic ratio*, C_∞, is the square of the ratio of the actual root-mean-square length to the $n^{1/2}\ell$:

$$C_\infty = <r_o^2>/n\ell^2. \qquad (15.6)$$

The characteristic ratio takes into account both the steric hinderance and valence angle terms. It is a good indicator of the stiffness of the chain. Typical values of σ and C_∞ are listed in Table 15.2. It is clear that the stiffness decreases with increasing temperature and that large side groups increase the stiffness.

Hard sphere model

The degree of short-range order in an amorphous material can be characterized by a hard sphere model if the basic structure of an amorphous material is approximated by spheres. The density of packing of atoms around a reference atom is described by the number of atom centers per volume that lie in a spherical shell of thickness, dr, and radius about the reference atom. In a hard sphere model, the number, n, of neighboring spheres with centers between r and dr is measured as a function of r.

Table 15.2. Typical values of steric hindrance and the characteristic ratio

Polymer	Temperature, °C	σ	C_∞
Polyethylene	140	1.8	6.8
Polypropylene	140	1.6	5.2
PVC	25	1.8	6.7
Polystyrene	25	2.3	10.8
Polystyrene	70	2.1	9.2
PMMA	25	2.1	8.6
PMMA	72	1.8	6.6

Source: From R. J. Young and P. A. Lovell, *Introduction to Polymers*, 2nd ed. (London: Chapman & Hall, 1991), p. 160.

For example, consider the two-dimensional array in Figure 15.3. For a two-dimensional material, the function is

$$g(r) = 1/(2\pi r)(dn/dr). \tag{15.7}$$

For three dimensions, the appropriate function is

$$g(r) = 1/(4\pi r^2)(dn/dr). \tag{15.8}$$

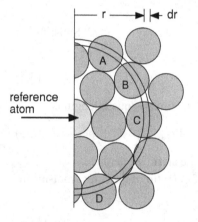

15.3. Two-dimensional distribution of hard spheres. The function $g(r)$ is the number of spheres whose centers are at a distance between r and $r + dr$ from the reference sphere. In this case the centers of spheres A, B, C, and D lie between r and $r + dr$ from sphere 1 so $g(r) = 4$ at the distance r.

In the hard sphere model, atoms may not overlap so $g(r) = 0$ for $r \leq D$, where D is the atom diameter. For completely random packing, $g(r)$ is constant for $r \geq D$. In a crystalline material, $g(r) = 0$ except at discrete interatomic distances. Figure 15.4 shows a schematic plot of $g(r)$ versus r for random packing, a crystal,

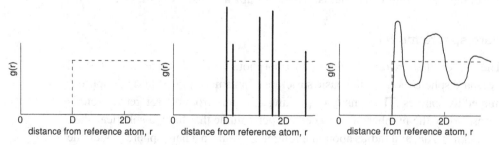

15.4. Schematic plot of $g(r)$ as a function of r for a random material, a bcc crystal, and a glass with short-range crystal.

AMORPHOUS MATERIALS

and an amorphous material. The value of $g(r)$ for the amorphous material is intermediate. There is some order at short range, but this disappears for $r > 3D$.

Voronoi cells

A simulation of the three-dimensional structure of an amorphous material can be made by establishing a random set of atom or molecule centers such that none overlap. Planes that bisect straight lines between near neighbors form Voronoi polyhedra or Wigner–Seitz cells. A two-dimensional construction is illustrated in Figure 15.5. This type of construction has been used to calculate the number of faces of three-dimensional polyhedra and the number of edges per face, using models of solid, liquid, and gaseous states. Finney calculated that, for a random solid, the cells have an average of 14.26 faces and the faces have an average of 5.158 edges. Similar calculations for a gas yield slightly higher numbers.

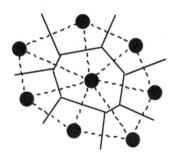

15.5. Construction of two-dimensional Voronoi cells by bisecting lines that connect near neighbors.

Silicate glasses

Zachariasen and co-workers[*] formulated four requirements for formation of glass from oxides. The requirements are:

1. Oxygen atoms are linked to no more than two anions.
2. The coordination number of the glass-forming anion is small.
3. The polyhedra formed by oxygens share corners, not edges or faces.
4. Polyhedra are linked in a three-dimensional network.

Only oxides A_2O_3, AO_2, and A_2O_5 satisfy these requirements. Triangular coordination with A_2O_3 and tetrahedral coordination with AO_2 and A_2O_5 are possible.

The basic structural units of silicate glasses are tetrahedra with Si^{+4} in the center bonded covalently to O^{-2} at each corner. In pure silica all corner oxygen ions are shared by two tetrahedra (Figure 15.6). The result is a covalently bonded glass with a very high viscosity. As other oxides are added, not all of the oxygen ions share two corners. This lowers the viscosity.

---

[*] W. H. Zachariasen, *J. Amer. Chem. Soc.* 84 (1932): 3841.

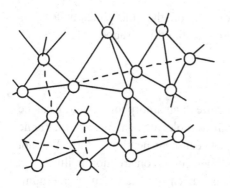

15.6. A silica glass is composed of tetrahedra with four O^{-2} surrounding a Si^{+4} at each center. Each O^{-2} is shared by two tetrahedra.

Typical glasses compositions are quite complex. A soda–lime glass may contain 72% SiO_2, 14% Na_2O, 12% CaO, and 3% MgO. The Na^+, Ca^{+2}, and Mg^{+2} bond ionically to some of the corner O^{-2} (Figure 15.7).

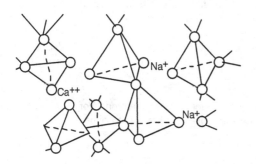

15.7. Commercial glasses contain alkali and alkaline earth ions, which substitute ionic bonds for the covalent bonds between tetrahedra.

Chemical composition

The chemical components of silicate glasses can be divided into three groups:

1. *Glass formers* include SiO_2 and B_2O_3. In a pure B_2O_3 glass the boron ions are in the center of a circle surrounded by three oxygen ions, each of which is shared with another triangle.
2. *Modifiers* such as Na_2O, K_2O, CaO, and MgO ionically bond with corners of the silica tetrahedra thus causing "nonbridging" oxygen ions. They tend to decrease the overall bond strength and thereby lower the viscosity.
3. *Intermediates* such as Al_2O_3 and PbO do not form glasses themselves but may join in the silica network. When Al_2O_3 is added to glass, some of the Al^{+3} ions act as intermediates, occupying centers of the tetrahedra, and some act as modifiers. Finally, some oxides such as B_2O_3 are glass formers. Pure B_2O_3 can form a glass, with triangles as the basic structural unit with three O^{-2} surrounding and covalently bonded to B^{+3}. This is possible because of the very small size of the B^{+3} ion.

Bridging versus nonbridging oxygen ions

In a silicate glass, each monovalent cation (Na^+ or K^+) contributes one unbonded O corner and each divalent cation (Ca^{+2} or Mg^{+2}) contributes two. The properties

AMORPHOUS MATERIALS

change as the number of unbonded corners increases. The number of unbonded corners per tetrahedron is $2N/S$, where N is the mole fraction of modifiers (each contributing two oxygens) and S is the mole fraction of SiO_2. The number of bridging oxygen ions per tetrahedron, Y, is

$$Y = 4 - 2N/S. \qquad (15.9)$$

Substituting $N = 1 - S$, Y can be expressed as

$$Y = 6 - 2/S. \qquad (15.10)$$

EXAMPLE 15.1. A soda–lime glass contains 73% SiO_2, 13% Na_2O, 11% CaO, and 3% CaO. Find Y and the O/Si ratio.

SOLUTION: There are 73/44 moles of SiO_2, 13/62 moles of Na_2O, 11/56 moles of CaO, and 3/24 moles of MgO so $Y = (13/62 + 11/56 + 3/24)/(73/44) = 0.64$ unbonded corners per tetrahedron. The ratio of O/Si is $4 - 0.64 = 3.36$.

Glass viscosity

Figure 15.8 shows the temperature dependence of several glass compositions. Several temperatures are identified by viscosity: the working temperature of a glass by 10^3 Pa · s, the softening point by 4×10^6 Pa · s, the anneal point by 2.5×10^{12} Pa · s, and the strain point by 4×10^{13} Pa · s. Glass objects are usually shaped

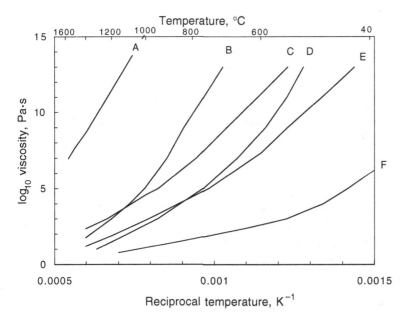

15.8. The temperature dependence of viscosity for several glasses. The compositions of the glasses are: A, $SiO_2 + 0.12\%$ H_2O; B, alumino–silicate glass 64% SiO_2, 4.5% B_2O_3, 10.4% Al_2O_3, 8.9% CaO, 10.2% MgO, 1.3% Na_2O, 0.7% K_2O; C, borosilicate glass (Pyrex) 81% SiO_2, 13% B_2O_3, 4% Na_2O, 2% Al_2O_3; D, soda–lime glass 70% SiO_2, 21% Na_2O, 9% CaO; E, alkali–lead glass, 77% SiO_2, 9% Na_2O, 1% CaO, 5% K_2O, 8% PbO; F, B_2O_3.

at or near the working temperature. The weight of glass objects will cause appreciable creep above the softening point. Stress relief occurs at the annealing point in 15 min. Figure 15.8 shows that the temperature dependence of viscosity is not well described by an Arrhenius equation. The Vogel–Fulcher–Tammann equation,

$$\eta = K\exp[E/(T - T_o)], \qquad (15.11)$$

describes the temperature dependence better.

The viscosities of glasses increase with the fraction of bridging oxygens. Figure 15.9 shows the increase of η with Y.

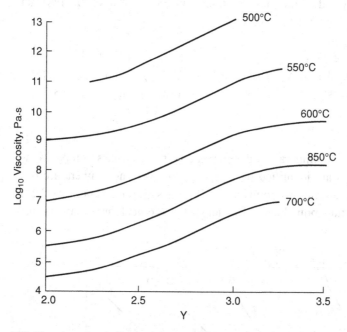

15.9. The viscosity of silicate glasses drops with decreasing O/Si ratios. The data from H. J. L. Trap and J. M. Stevels, *Glastech Ber*. 6 (1959): V 131, are for equal molar additions of Na_2O, K_2O, CaO, SrO, and BaO.

Thermal shock

Glasses, like most ceramics, are susceptible to fracturing under stresses caused by temperature gradients. Internal stresses in a material arise when there are different temperature changes in adjacent regions. In the absence of stress, a temperature change causes a fractional dimensional change, $\Delta L/L = \varepsilon = \alpha \Delta T$. Under stress, the total strain is

$$\varepsilon_x = \alpha \Delta T + (1/E)[\sigma_x - \nu(\sigma_y + \sigma_z)]. \qquad (15.12)$$

When two regions, A and B, are in intimate contact they must undergo the same strains ($\varepsilon_{xA} = \varepsilon_{xB}$). If there is a temperature difference, $\Delta T = T_A - T_B$, between the two regions,

$$\alpha \Delta T + (1/E)[\sigma_{xA} - \sigma_{xB} + \nu(\sigma_{yA} + \sigma_{zA} - \sigma_{yB} - \sigma_{zB})] = 0. \qquad (15.13)$$

EXAMPLE 15.2. The temperature of the inside wall of a tube is 200 °C and the outside wall temperature is 40 °C. Calculate the stresses at the outside of the wall if the tube is made from a glass having a coefficient of thermal expansion of $\alpha = 8 \times 10^{-6}/°C$, an elastic modulus of 60 GPa, and a Poisson's ratio of 0.3.

SOLUTION: Let x, y, and z be the axial, hoop, and radial directions. The stress normal to the tube wall is $\sigma_z = 0$ and symmetry requires that $\sigma_y = \sigma_x$. Let the reference position be the mid-wall where $T = 120$ °C. ΔT at the outside is $40 - 120 = -80$ °C. The strains ε_x and ε_y must be zero relative to the mid-wall. Substituting into Hooke's law, $0 = \alpha \Delta T + (1/E)(\sigma_x + \nu(\sigma_y + \sigma_z))$, $\alpha \Delta T + (1 - \nu)\sigma_x/E = 0$, so $\sigma_x = \alpha E \Delta T/(1 - \nu)$. $\sigma_x = (8 \times 10^{-6}/°C)(80\,°C)(60\,\text{GPa})/0.7 = 54$ MPa.

In general, the stresses reached will be proportional to α, $E/(1-\nu)$, and ΔT. The parameter,

$$R_1 = \sigma_f(1 - \nu)/(E\alpha), \tag{15.14}$$

describes the relative susceptibility to thermal shock. A different thermal shock parameter,

$$R_2 = K_{Ic}/(E\alpha), \tag{15.15}$$

is based on the fracture toughness. If the length of preexisting cracks is constant, these are equivalent because σ_f is proportional to the fracture toughness, K_{IC}. Thermal conductivity has some influence on susceptibility to thermal shock. A high thermal conductivity reduces the ΔT term.

Because E does not differ greatly among the various grades of glass, differences of thermal shock resistance depend primarily on differences in thermal expansion.

Thermal expansion

Additions of alkali and alkaline earth elements are used in glasses to lower the viscosity at temperatures low enough for the glass to be economically formed into useful shapes. However, these additions also raise the coefficient of thermal expansion. Figure 15.10 shows the relation between the coefficient of thermal expansion and the temperature at which the viscosity is 10^7 Pa · s, which is considered a temperature for forming.

Vycor

Vycor was developed by Corning Incorporated to provide a way around the problem of the difficulty in forming glasses of low thermal conductivity. The viscosity of the starting composition (62.7% SiO_2, 26.9% B_2O_3, 6.6% Na_2O, and 3.5% Al_2O_3) is low enough for the glass to be shaped at a reasonable temperature. After shaping, the glass is heat treated between 500 and 750 °C. A spinodal

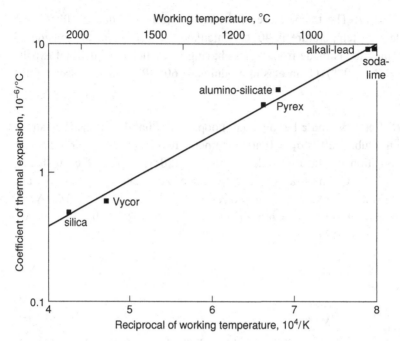

15.10. The relation between the coefficient of thermal expansion and the temperature at which the viscosity is 10^3 Pa · s.

reaction occurs during the heat treatment that separates the glass into two phases, one containing 96% SiO_2 and the other with most of the other components. The impurity-rich phase is removed by acid etching, leaving a silica-rich glass with about 28% porosity. This can either be used as a filter or reheated to allow sintering to produce a fully dense product.

Devitrification

If a glass is held for a long period at an elevated temperature it may start to crystallize or *devitrify*. Devitrification of fused quartz (silica glass) to cristobolite is slow. Nucleation is usually at a free surface and is often stimulated by contamination from alkali ions such as sodium. The rate of growth of cristabolite is increased by oxygen and water vapor. With surface contamination, devitrification of fused quartz may occur at temperatures as low as 1000 °C. However, if the surface is clean it rarely occurs below 1150 °C.

Glasses may intentionally be made to crystallize. David Stookey at Corning Incorporated discovered a way of producing fine-grained ceramics by crystallizing a glass. This process has been commercialized as Pyroceram and CorningWare. It involves a lithia–alumina silicate with TiO_2 added as a nucleating agent. Processing involves forming objects to its final shape at an elevated temperature, heat treating at a lower temperature to allow nucleation of crystals, and then reheating to a higher temperature for growth. This allows glass-forming processes to be used to obtain the final shapes and produces a final product

AMORPHOUS MATERIALS

that is resistant to thermal shock because of a very low thermal expansion coefficient.

Delayed fracture

Glass that has been under stress for a period of time may fracture suddenly. Such delayed fracture is not common in metals (except in cases of hydrogen embrittlement of steels) but sometimes does occur in polymers. It is often called *static fatigue*. The phenomenon is sensitive to temperature and prior abrasion of the surface. Most important, it is very sensitive to environment. Cracking is much more rapid with exposure to water than if the glass is kept dry (Figure 15.11) because water breaks the Si–O–Si bonds by the reaction $-$Si–O–Si$-$ $+$ H_2O $\to$ Si–OH $+$ HO–Si.

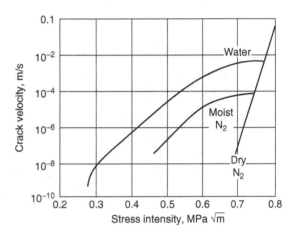

15.11. The effect of environment on crack velocity in a silicate glass under load. Reprinted with permission of ASM International® from *Engineering Materials Handbook*, vol. 4, *Ceramics and Glasses* (Materials Park, OH: ASM, 1991), p. 658. All rights reserved. www.asminternational.org.

Other inorganic glasses

Borax, B_2O_3, forms a glass in which the basic structural elements are triangles with boron at the center surrounded and covalently bonded to three oxygen atoms. Each of the oxygen atoms is shared by three triangles, as shown in Figure 15.12.

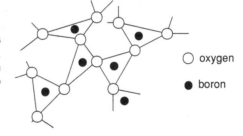

15.12. Borax glass. Each boron atom is covalently bonded to three oxygen atoms, which form a triangle around the boron atom. Each oxygen atom is shared by two triangles.

Chalcogenide* glasses consist of long Se (or Te) chains bonded with Ge or As. In these the basic structural units are chains that are cross-linked by As or Ge. The structure of molten pure selenium and pure tellurium consists of

---

* Chalcogens are O, S, Se, and Te.

long chain molecules. These form glasses if cooled rapidly. However, they will crystallize if heated between the glass transformation temperature and the melting point. Small amounts of arsenic or germanium will form a network and prevent crystallization. Figure 15.13 shows the structure schematically. Such glasses are useful in xerography.

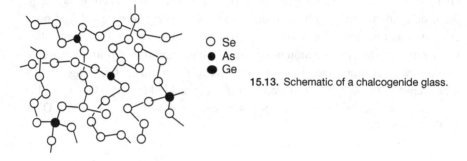

15.13. Schematic of a chalcogenide glass.

Metal glasses

Crystallization can be prevented in certain alloys if they are cooled rapidly enough. In these cases metallic glasses will form. Compositions of glass-forming alloys have several common features:

1. The equilibrium diagrams consist of two or more phases. Redistribution of the elements by diffusion is necessary for crystallization.
2. The compositions correspond to deep wells in the equilibrium diagram so the liquid phase is stable at low temperatures where diffusion is slow. See Figure 15.14.

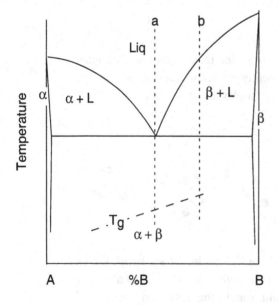

15.14. Schematic phase diagram. Composition **a** is more likely to form a glass than composition **b**, because of the much lower temperature at which crystallization can start.

3. The compositions usually have large amounts of small metalloids like B, C, P, Be, and Si.

Table 15.3. Properties of Vitreloy 1

Density	5.9 mg/m^3
Young's modulus	95 GPa
Shear modulus	35 GPa
YS	1.9 GPa
K_{IC}	55 MPa$\sqrt{m}$
T_g	625 K
Endurance limit/UTS	≈ 0.03

Source: From R. Zallen, *Physics of Amorphous Solids* (New York: Wiley, 1983), p. 6, table 1.1.

The formation of metal glasses by rapid cooling was first reported by Paul Duwez and co-workers in the 1960s.[*] They achieved cooling rates of thousands of degrees per second by shooting a fine stream of liquid metal onto a water-cooled copper drum. With the early compositions, cooling rates of about 10^5 K/s were necessary to prevent crystallization. This limited alloys to thin ribbons or wires.

More recently magnesium-base, iron-base, and zirconium–titanium-base alloys have been developed that do not require such rapid cooling. In 1992, W. L. Johnson and co-workers developed the first commercial alloy available in bulk form: Vitreloy 1, which contains 41.2 a/o Zr, 13.8 a/o Ti, 12.5 a/o Cu, 10 a/o Ni, and 22.5 a/o Be. The critical cooling rate for this alloy is about 1 K/s so glassy parts can be made with dimensions of several centimeters. Its properties are given in Table 15.3.

Metal glasses have very high yield strengths. This permits very high elastic strains and therefore storage of a large amount of elastic energy. The high ratio of yield strength to Young's modulus permits elastic strains of $1.9/95 = 2\%$. The tensile stress–strain curves of glassy metals show almost no work hardening. Tensile tests are characterized by serrated stress–strain curves resulting from sudden bursts of deformation localized in narrow shear bands with abrupt load drops. The net effect is that the total plastic strain is quite limited. The localization can be explained partially by the lack of work hardening. The formation of free volume and adiabatic heating have been offered as explanations. The fracture toughness is very high but the fatigue strength is very low. The ratio of endurance limit to yield strength of 0.03 is very much lower than the ratios of 0.3 to 0.5 typical of crystalline metals.

There are two principal uses for glassy metals. Because metal glasses have no barriers for domain wall movement they are excellent soft magnetic materials. Thin ribbons have been used for transformer cores since the 1960s. Metallic glasses have very good corrosion resistance and very low damping.

The other major application is based on the large amount of elastic energy that can be stored. The very high yield strengths typical of metallic glasses permit very high elastic strains and therefore storage of a large amount of elastic

---

[*] W. Klement, R. H. Willens, and P. Duwez, *Nature* 187 (1960): 869.

energy. Commercial use has been made of metallic glass in the heads of golf clubs (Figure 15.15). The great capacity to store elastic energy has permitted longer drives.

15.15. Golf club heads of Vitreloy 1. Driver at the left and iron at the right. Courtesy of Otis Buchanan, Liquidmetal Technologies, Lake Forest, CA.

NOTE OF INTEREST

There is a popular myth that the panes of stained glass windows in the very old European cathedrals are usually thicker at the bottom than at the top because the glass has crept over the centuries under its self-weight. If this were true, there would be large air gaps at the tops of the panes. The true explanation is that the glass varied in thickness when it was installed. Until the nineteenth century, sheet glass was made by spinning a hot viscous glob on a rod. Centrifugal force caused the glob to form into a disc, as shown in Figure 15.16. The disc was thicker near the center than at the edges, so panes cut from it had a thickness variation. A good artisan would naturally install a pane with the thicker section at the bottom.

15.16. Before 1800, plate glass was made by spinning a glob of hot glass on the end of a rod, allowing centrifugal force to form a disc about 1.2 m in diameter. Panes were cut from the disc. Courtesy of Broadfield House Glass Museum, Kingsford, UK.

REFERENCES

R. J. Charles. *Progress in Ceramic Science* 1 (1961): 1–38.
R. H. Doremus. *Glass Science*, 2nd ed. New York: Wiley, 1994.
R. W. Douglas. *Progress in Ceramic Science* 1 (1961): 200–23.
R. J. Young and P. A. Lovell. *Introduction to Polymers*, 2nd ed. London: Chapman & Hall, 1991.
J. M. Ziman. *Models of Disorder*. Cambridge, U.K.: Cambridge Univ. Press, 1970.

PROBLEMS

1. Describe the shape of the Voroni cell where the centers of cells are arranged in a body-centered cubic arrangement. How many faces does it have?

2. What fraction of the oxygen ions are unbonded in a class that has a composition of 70% SiO_2, 15% Na_2O, 11% CaO, and 4% MgO?

3. What is the root-mean-square length of a molecule of polypropylene of MW = 5000? Compare this to the contour length of the molecule.

4. Plot $g(r)$ versus r for a bcc crystal out to $4\,r$.

5. Plot $g(r)$ versus r for a two-dimensional crystal with hcp packing.

6. Predict the viscosities of the starting compositions of Vycor glass (62.7% SiO_2, 26.9% B_2O_3, 6.6% Na_2O, and 3.5% Al_2O_3) at 700 °C.

7. Calculate the maximum amount of elastic energy that can be stored per kg of:
   A. High strength steel ($YS = 1.0$ GPa, E = 207 GPa, $\rho = 7.87$ kg/m^3).
   B. Vitreloy 1 (properties given in Table 15.3).

8. Assume that if Figure 15.8 is extrapolated to 200 °C, the viscosity of the lead–alkali glass is 10^{18} Pa · s. Using this viscosity, calculate the strain that would occur in five centuries under a stress of 0.007 Pa. Assume that the stress is caused by the weight of a panel of glass 1/3 m high. Note that $\varepsilon Y = \eta' \sigma$, where $\eta' = \sqrt{3}\eta$.

16 Liquid Crystals

The structures of liquid crystals are intermediate between the amorphous and crystalline states. They have some short-range orientational order. Some also have positional order. Thousands of organic compounds exhibit liquid crystal structures. Most have molecules that are very long and thin, but some have molecules that are flat and pancake shaped. Many compounds may exist in more than one liquid crystalline state. Transitions from one state to another may be *thermotropic* (caused by temperature change) or *lyotropic* (caused by change of solute concentration).

Types of liquid crystals

Liquid crystals may be classified as *nematic, cholesteric, smectic,* or *columnar*. Nematic liquid crystals are characterized only by orientational order. Molecules tend to be aligned with a director, as illustrated in Figure 16.1.

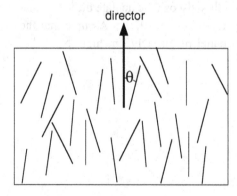

16.1. Nematic order in which there is statistical alignment with a director.

In cholesteric structures there is also alignment, but the direction of alignment rotates on a screw axis normal to the direction of alignment (Figure 16.2). This spiraling is responsible for unique optical properties.

Smectic liquid crystals have both positional and orientational order. The molecules are grouped into layers. In smectic A structures, the molecules tend

LIQUID CRYSTALS

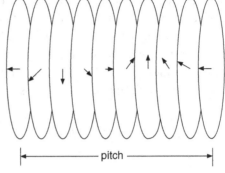

16.2. Order in a cholesteric liquid crystal. The alignment direction spirals about a screw axis normal to the alignment direction.

to be oriented normal to the layers, whereas in smectic C structures the direction of alignment is tilted away from the normal to the layers. These are shown in Figure 16.3.

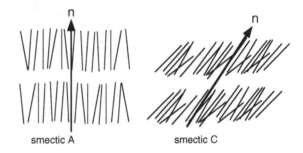

16.3. Alignment in smectic A and B.

Columnar liquid crystals consist of flat, disc-shaped molecules aligned in columns. These columns may be arranged in a hexagonal pattern as illustrated in Figure 16.4.

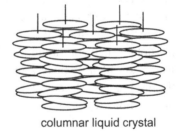

16.4. In columnar liquid crystals flat molecules are aligned in columns.

Orientational order parameter

The degree of orientational order can be described in statistical terms by the angular deviations of molecular orientations, θ, from the director. The parameter

$$s = <3\cos^2\theta - 1>/2 \tag{16.1}$$

is based on the average value of $(3\cos^2\theta - 1)/2$ rather than on the average value of θ. The reason for using this parameter is that it better represents the statistical distribution of possible orientations. If all orientations were at $\theta = 0$ to the director, $s = 1$, and if all of the orientations were at $90°$, $s = -1/2$. The value of s for randomly oriented molecules can be found by considering a hemisphere of all possible orientations (Figure 16.5). The fraction of possible orientations between angles, θ, and $\theta + d\theta$, to a reference direction is $\sin\theta d\theta/(\pi/2)$. The average value of the quantity $s = (3\cos^2\theta - 1)/2$ for randomly oriented material is

$$s = (1/2)(1/2\pi)\int_0^{\pi/2}(3\cos^2\theta - 1)\sin\theta d\theta = 0. \tag{16.2}$$

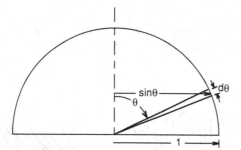

16.5. The fraction of the area of a hemisphere between θ and $\theta + d\theta$ equals $\sin\theta d\theta/(\pi/2)$.

Disclinations

Liquid crystals may have line defects called *disclinations*. The name comes from *discontinuity* and *inclination*. The director rotates about a line normal to the disclination. The strength of a disclination, S, is defined by

$$S = \Omega/2\pi, \tag{16.3}$$

where Ω is the angular rotation of the director in a circuit about the disclination. If the direction of rotation is the same as the direction of the circuit, S is positive. If the direction of rotation is the opposite of the direction of the circuit, S is negative. Figure 16.6 illustrates three possibilities. The configurations in Figure 16.6 are possible in nematic liquid crystals. It should be noted that they are similar to whorls that identify fingerprints. The director is in a plane that is perpendicular to the disclination. Because the energy of a disclination is proportional to S^2, only disclinations with low values of S are common.

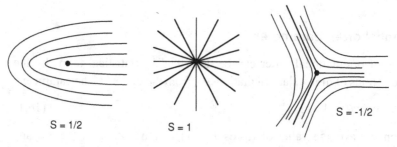

16.6. Several examples of disclinations viewed parallel to the line of the disclination.

LIQUID CRYSTALS

Lyotropic liquid crystals

Lyotropic liquid crystals form in solutions of polar molecules such as soap in water. One end of the molecule is hydrophilic and the other end is hydrophobic. The molecules are aligned such that the hydrophilic end is exposed to water and the hydrophobic end is shielded from the water. There are several forms. The molecules may be arranged in lamellae or spherical units (Figure 16.7). The spherical units tend to be arranged in body-centered cubic arrays. The lamellae may be flat or rolled up to form columns that are arranged in hexagonal patterns.

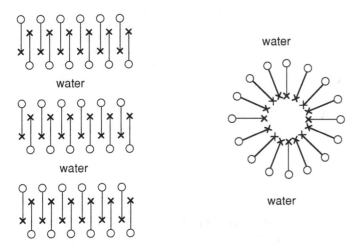

16.7. Arrangements of polar molecules in lyotropic liquid crystals. Open circles indicate hydrophilic ends and x's indicate hydrophobic ends.

Temperature and concentration effects

Liquid crystals are stable only over limited temperature ranges. The degree of orientation decreases with increasing temperature. There is a critical temperature, T_c, at which s drops to zero, as shown in Figure 16.8. This is analogous to Curie temperature, T_c in ferromagnetic materials, at which the material ceases to be ferromagnetic.

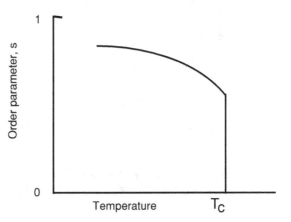

16.8. The order parameter, s, decreases with increasing temperatures, abruptly dropping to zero at the critical temperature for a nematic.

The order parameter is both temperature and pressure sensitive. The parameter, s, increases with pressure and decreases with temperature. Figure 16.9 shows that for the same degree of order,

$$d \ln T / d \ln V = -4. \tag{16.4}$$

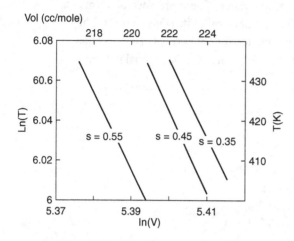

16.9. The effect of temperature and volume on the order parameter, s, in polyacrylic acid. The order increases with pressure and decreases with temperature. The straight lines indicate the validity of Equation 16.4. Data from J. R. McColl and C. S. Shih, *Phys. Rev. Letters* 29 (1972): 85.

Phase changes

One compound may exist in more than one liquid crystal form. The structure of *cholestryl myristate* changes from solid to smectic A at 71 °C, to chiral nematic at 79 °C, and finally to an isotropic liquid at 85 °C. Figure 16.10 shows the transitions in *p*-methoxybenzoic acid. Note that the order increases with increased pressure and lower temperature. Liquid crystal phases are more densely packed than the

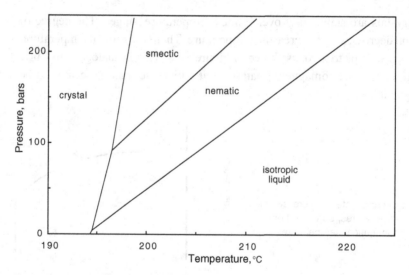

16.10. Phase diagram for *p*-methoxybenzoic acid. Data from S. Chandrasekhar, S. Ramaseshan, A. S. Reshamwala, B. K. Sadashiva, R. Sashudahar, and V. Surendranath, *Proc. Internat. Liquid Crystal Conf.* (1973): 117.

isotropic liquid, though volume changes between liquid phases are small. For the nematic to isotropic liquid state, ΔV is 2 to 10% of that for crystal to isotropic liquid. The latent heats of transition between liquid crystal phases are also very small. For the nematic to isotropic liquid state, ΔH is between 1 and 5% of that for crystal to isotropic liquid.

Optical response

Liquid crystalline materials may appear cloudy. This is because the index of refraction of polarized light depends on the angle between the angle of polarization and the director. This is represented schematically in Figure 16.11. In a nematic material, there will normally be regions in which the director is oriented differently.

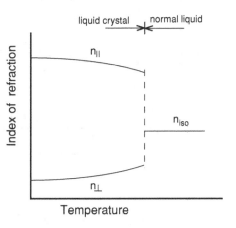

16.11. The index of refraction depends on the angle between the director and the polarization.

In nematic single crystals, the molecules have some degree of polarization. The charge is likely to be positive at one end and negative at the other. Because of this an electric field will cause the director to rotate into alignment. Rotation of the direction of polarization is the basis of liquid crystal displays. They can be made to be opaque or transparent to polarized light.

Temperature changes also can produce color changes in chloresteric liquid crystals, as illustrated in Figure 16.12.

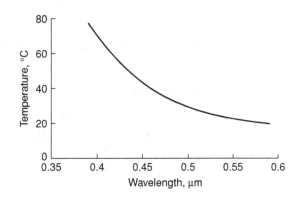

16.12. Temperature dependence of wavelength scattering for 20% chlosteryl acetate and 80% chlosteryl nonanate. Data from J. L. Fergason, N. N. Goldberg, and R. J. Nadalin, *Mol. Cryst.* 1 (1966): 309.

Liquid crystal displays

The most important applications of liquid crystals are as displays on all sorts of electronic devices. The principle of these displays is illustrated in Figure 16.13. A nematic liquid crystal is placed between two polarized plates of a polymer. Previous rubbing of the inner surfaces of these plates causes the molecules to align with the surfaces. The direction of alignment on the top plate differs by 90° from that of the bottom plate. Polarization of the plates prevents light from being transmitted. If a field is applied to the plates, the molecules tend to align with the field, allowing light to pass.

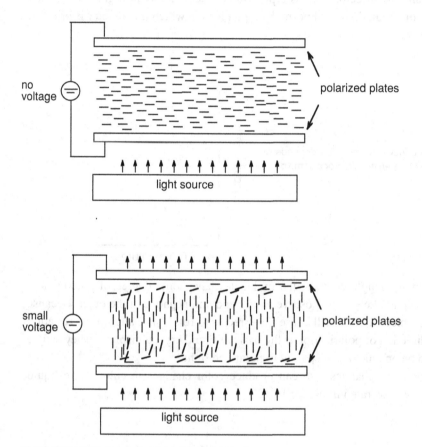

16.13. Schematic drawing of a liquid crystal display.

NOTE OF INTEREST

Freidrich Reinirzer first reported liquid crystals. He found that a crystalline cholesterol ester first melted at 145.5 °C into a cloudy liquid that subsequently melted into a clear liquid at 178.5 °C. The term *liquid crystal* was first coined in 1890 by Lehman.

LIQUID CRYSTALS

REFERENCES

P. M. Chaikin and T. C. Lubensky. *Principles of Condensed Matter Physics.* Cambridge, U.K.: Cambridge Univ. Press, 1995.

S. Chandrasekhar. *Liquid Crystals.* Cambridge, U.K.: Cambridge Univ. Press, 1992.

P. J. Collins and M. Hird. *Introduction to Liquid Crystals Chemistry and Physics.* New York: Taylor & Francis, 1997.

PROBLEMS

1. Determine the value of S for the disclinations in Figure 16.14.

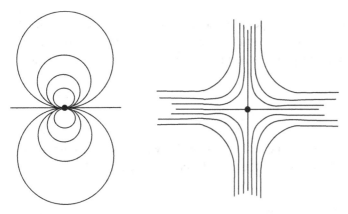

16.14. Two disclinations.

2. Sketch a disclination of strength $S = -3/2$.

3. By what fraction would the specific volume of polyacrylic acid have to change keep the same degree of order, s, as a change of temperature from 140 to 145 °C?

4. Determine the value of s if all of the molecules were oriented at 30° to the director.

17 Molecular Morphology

The shape of molecules and the types of bonding between them have significant effects on properties of materials. Examples include the wide variety of molecular shapes of silicates and the planar structures of talc, mica, clay, molybdenum disulfide, and graphite.

Silicates

The basic structural units of silicates are tetrahedra with a silicon atom at the center surrounded by four oxygen atoms, as sketched in Figure 17.1.

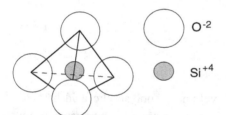

17.1. The basic tetrahedron of silicates has a Si^{+4} ion at the center surrounded by four O^{-2} ions on the corners.

The molecular structure depends on the oxygen-to-silicon ratio. If $O/Si = 2$, each oxygen is covalently bonded to two tetrahedra so a three-dimensional network is formed. At the other extreme, if the $O/Si = 4$, none of the four oxygen atoms is shared by another tetrahedra, and isolated molecules are formed. Figure 17.2 shows many of the possibilities.

For $O/Si = 4:1$ or $3.5:1$, isolated molecules are formed. An example is fosterite, Mg_2SiO_4. For $O/Si = 3:1$ or $2.75:1$, linear chains are formed. Asbestos is an example. The linear molecules are bonded to one another by weak ionic bonds. The result is a mineral that can be torn into fibers. For $O/Si = 2.5:1$, sheets are formed. One example is talc, $[Mg_3Si_4O_{10}(OH)_2]_2$. Its lubricity is due to the ease with which the covalently bonded sheets can slide over one another. There are several forms of mica. In muscovite, $[KAl_3Si_3O_{10}(OH)_2]_2$, and in phlogolite, $[KAlMg_3Si_3O_{10}(OH)_2]_2$, one of the aluminum atoms is in the center of a tetrahedron, and the others act as modifiers. The double sheets are weakly bonded to

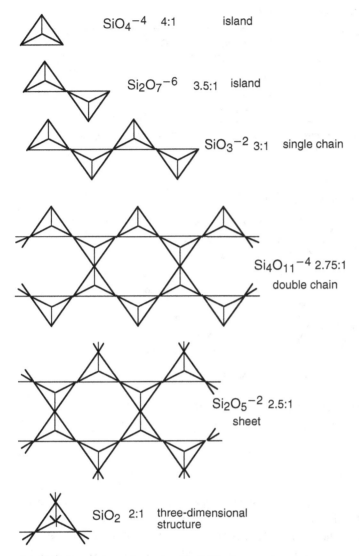

17.2. Silicate structures depend on the O/Si ratio.

one another (Figure 17.3). The very weak van der Waals bonding between these molecular sheets explains why mica cleaves so easily.

Clays are still another example of sheet structures. Kaolinite has the composition [$Al_2Si_2O_5(OH)_4$]. It consists of a $Si_2O_5^{-2}$ sheet bonded to an $Al_2(OH)_6$ sheet with two thirds of the $(OH)^-$ ions on one side of the $Al_2(OH)_6$ sheet replaced by unsatisfied oxygen ions on the $Si_2O_5^{-2}$ sheet. This creates a one-sided molecular structure that attracts water, which is responsible for the ability of wet clay to be shaped easily. The clay becomes rigid when it is dried.

If the O/Si ratio is less than 2.5, three-dimensional framework structures are formed. Silica, SiO_2, occurs in several crystalline forms. These are listed in Table 17.1. Silica can also occur as a glass (see Chapter 15). In all of these the basic structural unit is the tetrahedron.

Table 17.1. Forms of silica

Stable form	Temperature range, °C	Density
Low quartz	≤ 573	2.65
High quartz	573–867	
High tridymite	867–1470	2.26
High cristobalite	1470–1710	2.32

Source: From W. D. Kingery, *Introduction to Ceramics* (New York: Wiley, 1960). Reprinted with permission of John Wiley & Sons, Inc.

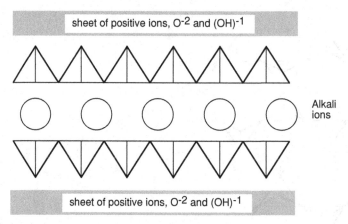

17.3. The structure of mica consists of two $Si_2O_5^{-2}$ sheets bonded together by alkali ions. There are layers of positive Al^{+3} and Mg^{+2} ions and negative O^{-2} and OH^{-1} ions on the outside of the double sheet.

Molybdenum disulfide

Molybdenum disulfide also forms sheet molecules. It consists of covalently bonded sheets with sulfur atoms on both sides of molybdenum atoms, as shown in Figure 17.4. These sheets are only weakly bonded. The ability of these sheets to slide over each other explains the lubricity of Mo_2S, which is used as a high-pressure solid lubricant in metal working.

17.4. Molybdenum disulfide is composed of covalently bonded sheets that are weakly bonded to each other.

MOLECULAR MORPHOLOGY

Table 17.2. Directional properties of graphite

Property	Perpendicular to c (∥ plane of sheet)	Parallel to c (⊥ plane of sheet)	
Electrical resistivity	2.5 – 5	3800	(ohm · m)
Thermal conductivity	398	2.2	W/m · K
Thermal expansion (20 °C)	slightly negative	25×10^{-6}	°C^{-1}
Elastic modulus	1060	35.5	GPa

Source: Data from H. O. Pierson, *Handbook of Carbon, Graphite, Diamond and Fullerenes* (Norwich, NY: Knovel, 2001).

Carbon: graphite

Graphite is composed of sheets of carbon atoms arranged in a hexagonal pattern (Figure 17.5). The bonding in the hexagonal sheets is like that in a benzene ring. Each carbon has two single bonds and a double bond. Only van der Waals bonds hold the sheets together. The ease with which sheets can slide over one another explains the lubricity of graphite. Because the double bond can move freely, the electrical conductivity in the plane of the sheet is very high, like a metal. The electrical and thermal conductivities perpendicular to the sheets are very low. Likewise, Young's modulus is very high in the planes of the sheet and very low perpendicular to them. The anisotropy of properties listed in Table 17.2 reflects the difference in bond strengths parallel and perpendicular to the sheets.

17.5. The structure of graphite.

Diamond

In diamond each carbon atom is covalently bonded to four others. Figure 17.6 shows the structure. The very strong bonding makes diamond the hardest material known. It has an extremely high Young's modulus (1.100 GPa) and a very low coefficient of thermal expansion (1×10^{-6}/K). Its very high thermal conductivity (1 to 2 kW/mK) makes it useful for dissipating heat. Its density (3.52 Mg/m^3) is considerably greater than that of graphite (2.25 Mg/m^3).

Hard amorphous carbon films may either be fully amorphous or contain tiny diamond crystallites. Hydrogen-free films may be deposited on surfaces from graphite by laser ablation or ion sputtering. Hydrogen-containing films are also possible.

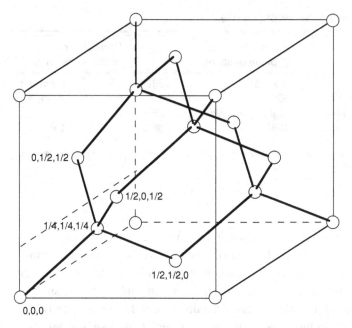

17.6. The crystal structure of diamond. Each carbon atom is covalently bonded to four others.

Carbon fibers

Carbon fibers are thin graphite ribbons. They are made by pyrolizing polymeric precursors. Typical precursors are PAN (polyacrylonitrile), pitch, rayon, or other polymers that have carbon–carbon backbones. Processing consists of several steps: stretching or spinning to align polymer chains, heating to stabilize the orientation, further heating to pyrolize, and still further heating to graphitize. The strengths and moduli are very high because they involve stretching carbon–carbon bonds. The level of the properties depends greatly on the nature of the precursor, its diameter, and the details of processing. Young's moduli of commercial carbon fibers vary from 200 to 700 GPa and tensile strengths vary from 2 to 7 GPa.

Carbon fibers are used in composite bonded with epoxy. By 2004, the market had grown to over 35,000 metric tons per year.

Fullerenes

Until 1985, the only known elemental forms of carbon were diamond, graphite, and amorphous carbon. Then Kroto et al.[*] announced the discovery of C_{60}, a spherical arrangement of carbon atoms in hexagons and pentagons, as shown in Figure 17.7. They called this form Buckminsterfullerene after the architect Buckminster Fuller, who developed the geodesic dome. The name for this type of carbon molecule has since been shortened to fullerene, but it is commonly called a buckyball. Since this first discovery, it has been found that fullerenes can be made in quantity from electrical arcs between graphite electrodes. About 75% of

---

[*] H. Kroto, J. Heath, S. O'Brien, R. Curl, and R. Smalley, *Nature* 318 (1985): 162–3.

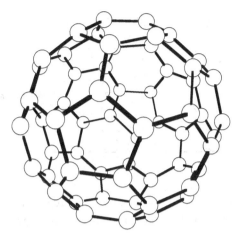

17.7. C_{60} buckyball. There are 60 carbon atoms arranged in hexagons and pentagons. The arrangement is the same as that on a soccer ball.

the fullerenes produced by arcs are C_{60}, 23% C_{80}, with the rest being even larger molecules. About 9000 fullerene compounds are known.

Nanotubes

A nanotube can be thought of as a hexagonal sheet of carbon atoms (graphene sheet), rolled up to make a cylinder and capped at the ends by a half of a buckyball, as illustrated in Figure 17.8. Tubes typically have diameters of about 1 nm. The diameter of the smallest nanotube corresponds to the diameter of the smallest buckyball (C_{60}.) The length-to-diameter ratio is typically about 10^4.

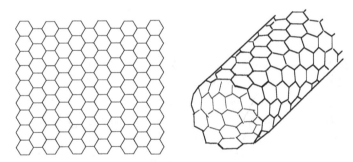

17.8. Singlewall nanotubes can be thought of as being made from rolled up chicken wire.

Nanotubes fall into three groups, depending on the chiral angle, θ, between the $<2\bar{1}\bar{1}0>$ direction of the hexagons and the tube axis (Figure 17.9). If $\theta = 0$, a zigzag nanotube results. If $\theta = 30°$, the nanotube is called an armchair. Chiral nanotubes are those for which $0 < \theta < 30°$. These develop twists. Nanotubes can have metallic conduction; others are semiconductors or insulators.

Rings can form from nanotubes, when the two ends join each other. Concentric multiwall nanotubes can form as well as single-thickness nanotubes.

There are a number of potential uses of fullerenes. One potential use of nanotubes is for field effect transistors. Nanotubes decorated with metal atoms have a great potential for hydrogen storage for fuel cells. A_3C_{60} compounds where

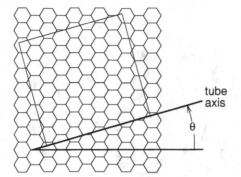

17.9. One characteristic of a nanotube is the angle between the tube axis and the crystallographic axes of the hexagons.

A is an alkali (K, Rb, Cs, Na) are superconductors. Sieves that allow biological compounds to pass through but not larger viruses have been suggested.

Zeolites

Zeolites are a class of porous minerals with Al^{+3} and Si^{+4}, tetrahedrally bonded with O^{-2} and monovalent and divalent cations like Na^+, K^+, Ca^{+2}, Mg^{+2}, and H_2O. Two examples are chabazite, $CaAl_2 Si_4O_{12} \cdot 2H_2O$, and natrolite, $Na_2Al_2Si_3O_{10} \cdot 2H_2O$. Each O^{-2} ion is coordinated with either two Si^{+4} ions or with one Si^{+4} ion and one Al^{+3} ion (never two Al^{+3} ions). There are channels through the zeolite crystals large enough to permit passage of small molecules. See Figure 17.10.

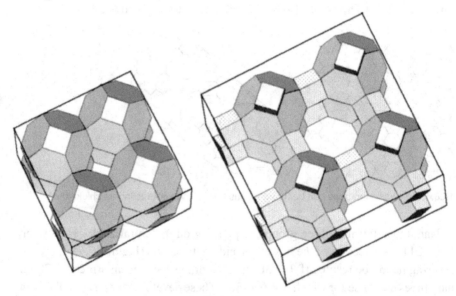

17.10. Structures of two zeolites. The Al^{+3}–Si^{+4} skeletons are represented. The holes in the zeolites range from 0.25 to 1 nm. From J. I. Gersten and F. N. Smith, *The Physics and Chemistry of Materials* (New York: Wiley, 2001). Reprinted with permission of John Wiley & Sons, Inc.

Heating drives off the H_2O without destroying the structure. The remaining material is hydrophilic and often used as a dessicant. Zeolites also facilitate ion exchange. One example is water softeners. When hard water containing Ca^{+2} ions is passed through on a zeolite on which Na^+ ions are adsorbed, two Na^+ ions exchange with Ca^{+2} ions. Another major use of zeolites is as catalysts.

NOTES OF INTEREST

1. The word *zeolite* comes from the Greek word meaning "weeping stone."
2. The discovery of C_{60} came about from an attempt by Kroto et al. to understand the adsorption spectra of interstellar space. Although they failed in this attempt, the discovery of C_{60} won them the Nobel Prize for chemistry in 1996.
3. Buckminster Fuller (1895–1983) was born in Milton, Massachusetts. He received many architectural awards including the Gold Medals of the American Institute of Architects and the Royal Institute of British Architects. He received 47 honorary doctorates, was awarded 27 U.S. patents, and wrote 25 books. He is best known for his introduction of the geodesic dome for buildings. There are now over 300,000 geodesic domes in the world.

REFERENCES

J.-B. Donnet. *Carbon Fibers*. New York: Marcel Dekker, 1998.

J. I. Gersten and F. W. Smith. *The Physics and Chemistry of Materials*. New York: Wiley, 2001.

W. D. Kingery. *Introduction to Ceramics*. New York: Wiley, 1960.

W. G. Moffatt, G. W. Pearsall, and J. Wulff. *Structure and Properties of Materials*, vol. I, *Structure*. New York: Wiley, 1964.

PROBLEMS

1. What are x and y in $(Si_xO_y)^{-m}$ for the following?
   A. single chain
   B. double chain
   C. sheet silicates
2. Predict the structures of these silicates.
   A. olivine, Mg_2SiO_4
   B. ultramarine, $Na_8(Al_6Si_6O_{24})Cl_2$
3. In graphite the distance between planes is 0.335 nm and the distance between carbon atoms in the hexagonal planes is 0.142 nm. Calculate the density of graphite.
4. How many faces does a C_{80} buckyball have? How many are hexagons and how many are pentagons?

18 Magnetic Behavior of Materials

Until about 200 years ago, magnetism was a mysterious phenomenon. The discovery of the magnetic effect in lodestone (Fe_3O_4) led to the first use of magnetism in compasses. When we speak of "magnetic behavior," we usually mean ferromagnetic behavior. All materials have some response to a magnetic field. Paramagnetic materials weakly repulse magnetic fields and diamagnetic materials weakly attract magnetic fields.

Ferromagnetism

In contrast, ferromagnetic materials very strongly attract magnetic fields. There are only a few ferromagnetic elements. The important ones are iron, nickel, and cobalt. A few rare earths are ferromagnetic at low temperatures. Atoms of other transition elements may be ferromagnetic in alloys or compounds where the distance between atoms is different than in the elemental state. These include the manganese alloys Cu_2MnAl, Cu_2MnSn, Ag_5MnAl, and $MnBi$. Table 18.1 lists a number of ferromagnetic elements, their Curie temperatures (temperature above which they cease to be ferromagnetic), and their saturation magnetizations.

Ferromagnetism arises because of an unbalance of electron spins in the 3d shell of the transition elements (the 4f shell for rare earths). The unbalanced spin causes a magnetic moment. In metals with valences of 1 or 3 (e.g., Cu or Al), each atom has an unbalance of spins, but the unbalance is random so there is no net effect. With the transition elements, the 3d and 4s energy bands overlap (Figure 18.1).

There are four important energy terms that affect ferromagnetic behavior:

1. *exchange energy*,
2. *magnetostatic energy*,
3. *magnetocrystalline energy*, and
4. *magnetostrictive energy*.

MAGNETIC BEHAVIOR OF MATERIALS

Table 18.1. Ferromagnetic materials

Metal	Curie temperature, °C	Saturation magnetization (tesla)
Iron	771	2.16
Cobalt	1121	1.87
Nickel	358	0.616
Gadolinium	20	8.
Terbium	−52	3.4
Dysprosium	−188	3.71
Ho	−253	3.87
MnBi	630	
MnSb	587	
$Fe_{80}B_{20}$	647	
Oxide		
Fe_3O_4	858	
$NiFe_2O_4$	858	
$MnFe_2O_4$	573	
CrO_2	386	
EuO	69	

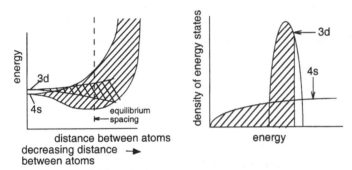

18.1. For the transition elements the 3d and 4s energy levels overlap.

Exchange energy

For some transition metals, the total energy is lowered in a magnetic field if one half of the 3d band is completely full, causing an unbalance of electron spins, as shown schematically in Figure 18.2. This results in a strong magnetic effect. In ferromagnetic materials, the field caused by neighboring atoms is strong enough to cause this shift.

This lowering of energy caused by alignment of the unbalanced spins with that of the neighboring atoms is called the exchange energy. It depends on the interatomic distance. For example, bcc iron is ferromagnetic but fcc iron is not. Figure 18.3 illustrates this. Figure 18.4 shows how the maximum number of unbalanced spins per atom (number of Bohr magnetons) depends on the number of 3d electrons.

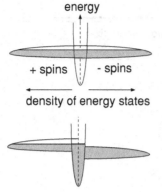

18.2. If one half of the 3d band is completely full and the other half partially full, there is a strong unbalance of electron spins causing a strong magnetic effect.

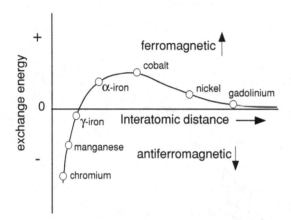

18.3. Dependence of exchange energy on atomic separation.

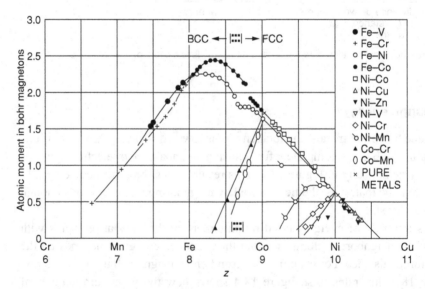

18.4. Variation of average atomic moment with the number of 3d plus 4s electrons in binary alloys of transition metals. From R. Bozorth, *Ferromagnetism* (Piscataway, NJ: IEEE Press, 1993).

MAGNETIC BEHAVIOR OF MATERIALS

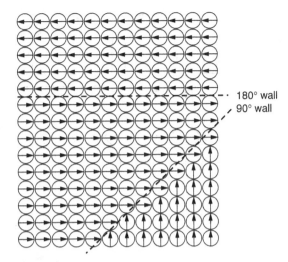

18.5. Ferromagnetic domains are regions in which unbalanced electron spins are aligned. Parts of three domains are indicated. The dashed lines are 180° and 90° domain walls. From W. F. Hosford, *Physical Metallurgy* (Boca Raton, FL: CRC Press, 2005), p. 445, figure 26.4.

The energy is minimized when neighboring atoms are magnetized in the same direction. This causes the formation of magnetic domains in which all of the neighboring atoms are magnetized in the same direction. These may contain 10^{15} atoms. Figure 18.5 schematically shows parts of three domains.

Magnetostatic energy

Incomplete magnetostatic circuits within the ferromagnetic material raise the total energy because the circuits must be completed externally (Figure 18.6). Horseshoe magnets attract iron is to complete their magnetostatic circuits in iron (Figure 18.7). A typical domain structure is composed of domains that form complete circuits, as shown in Figure 18.8.

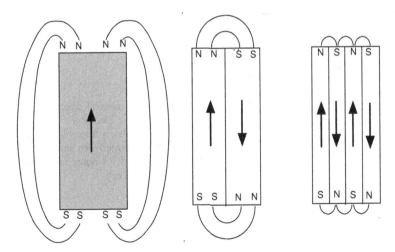

18.6. Incomplete magnetostatic circuits raise the energy.

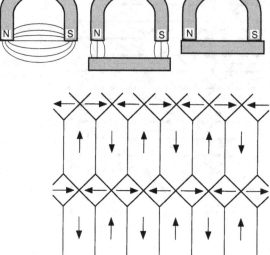

18.7. A horseshoe magnet attracts iron to complete a magnetostatic circuit.

18.8. Typical domain structure composed of complete magnetostatic circuits. From W. F. Hosford, *Physical Metallurgy* (Boca Raton, FL: CRC Press, 2005), p. 447, figure 24.5.

When the there are equal numbers of domains aligned in opposing directions, their magnetic fields cancel externally so the material appears not to be magnetized.

Magnetocrystalline energy

Each of the ferromagnetic materials has a specific crystallographic direction in which it is naturally magnetized. Figure 18.9 shows the B–H curves for iron

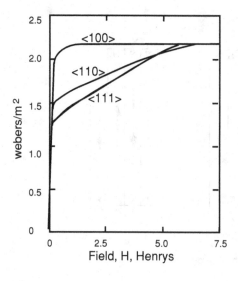

18.9. B–H curves for several directions in iron. After J. K. Stanley, *Electrical and Magnetic Properties of Metals* (Materials Park, OH: ASM, 1963). Reprinted with permission of ASM International.® All rights reserved. www.asminternational.org.

crystals of different orientations. For iron, the direction of easy magnetization is <100>, for nickel it is <111>, and for cobalt it is <0001>.

Magnetostrictive energy

Magnetization strains a material. In a cubic metal, the strain along a <100> direction is ε_{100} when the magnetic field is along that direction and the strain along a <111> direction is ε_{111} when the field is along the <111> direction. For iron $\varepsilon_{100} = 24 \times 10^{-6}$ and $\varepsilon_{111} = -23 \times 10^{-6}$, and for nickel $\varepsilon_{100} = -63.7 \times 10^{-6}$ and $\varepsilon_{111} = -29 \times 10^{-6}$. For randomly oriented material the value of ε is approximately

$$\varepsilon = 0.4\varepsilon_{100} + 0.6\varepsilon_{111}. \tag{18.1}$$

Stresses must arise if the magnetostrictive strains are prevented.

EXAMPLE 18.1. Calculate the energy associated with the magnetostriction near a 90° domain boundary of length L in iron.

SOLUTION: At a 90° domain boundary, there is a zero net strain in the [001] direction of domain A and the [010] direction of domain B (Figure 18.10). Therefore, there must be an elastic strain equal in magnitude and opposite in sign to the magnetostrictive strain. The compressive strain is ε_{100} and the associated stress is $E\varepsilon_{100}$, so the energy per volume is $(1/2)E\varepsilon_{100}^2$. The volume associated with this is $L^2 z/2$, where z is the dimension into the paper. Substituting $E_{100} = 130$ GPa and $\varepsilon_{100} = 24 \times 10^{-6}$ for iron, the energy is 780 kJ/m².

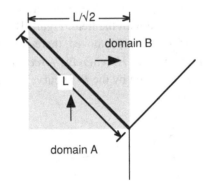

18.10. 90° domain boundary. The shaded area is the region in which the magnetostrictive strain must be compensated by elastic strains.

Physical units

Further discussion of magnetic behavior requires definition of some terms. The flux density or induction, B, in a material in a magnetic field is its response to the intensity of the magnetic field, H. The ratio of B to H is the permeability, μ:

$$B = \mu H. \tag{18.2}$$

In a vacuum,

$$B = \mu_o H, \tag{18.3}$$

where $\mu_o = 4\pi \times 10^{-7}$ Wb/(A/m). Equation 18.1 can be written as

$$B = \mu_o(H + M), \tag{18.4}$$

where M is the magnetization. Both the mks and cgs units to describe B, H, and μ are listed in Table 18.2 The intensity of the magnetic field or magnetizing force, H, is measured in A/m. The magnetic induction, B, is measured in teslas.

Table 18.2. Units

Quantity	Symbol	mks	cgs
Field strength	H	A/m	oersted = $4\pi \times 10^{-3}$ A/m
Flux	B	T = Wb/m^2	gauss = 10^4 T
Magnetic moment	M	A/m	emu/m^3 = 10^3 A/m
Permeability	μ	Wb/(A/m)	dimensionless

Note: The magnetic moment, M, caused by one unbalanced electron spin is called a Bohr magneton and has the value of 9.27×10^{-24} A/m^2.

The *B–H* curve

When a magnetic field is imposed on a ferromagnetic material, the domains most nearly aligned with the field will grow at the expense of the others, as illustrated in Figure 18.11. As they do, the material's magnetic induction will increase, as shown in Figure 18.12. At first, favorably aligned domains grow. Final induction occurs by rotation of the direction of magnetization out of the easy direction to be aligned with the field. Figure 18.13 shows an entire *B–H* curve.

If the field is removed, there is a *residual magnetization* or *remanence*, B_r. A reverse field, H_c (*coercive force*), is required to demagnetize the material. The area enclosed by the *B–H* curve (*hysteresis*) is the energy loss per cycle, and the

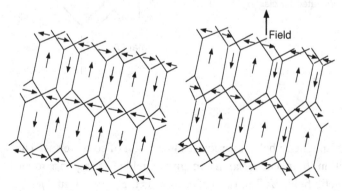

18.11. Imposition of an external field causes domains most nearly aligned with the field to grow at the expense of those antialigned.

MAGNETIC BEHAVIOR OF MATERIALS

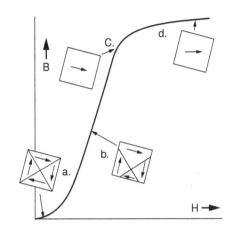

18.12. Magnetization of a material. Initially, magnetization increases by growth of favorably oriented domains. At high fields, the direction of magnetization rotates out of the easy direction.

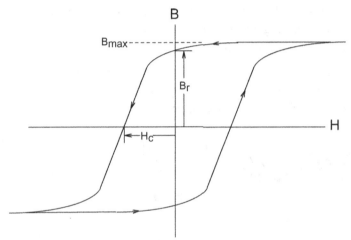

18.13. A typical B–H curve.

permeability is defined as $\mu = B/H$. The initial permeability, μ_o, and the maximum permeability, μ_{max}, are material properties.

Curie temperature

The Curie temperature is the temperature above which a material ceases to be ferromagnetic. Figure 18.14 shows the decrease of saturation magnetization, B_{max}, with temperature.

Bloch walls

The boundaries between domains are regions where there is a gradual change in the direction of magnetization. The width of these (perhaps 20 atoms) is a compromise between the magnetocrystalline and exchange energy terms. A wider boundary would require more atoms to be magnetized out of the direction of easy magnetization. The exchange energy is minimized if the boundary is very wide

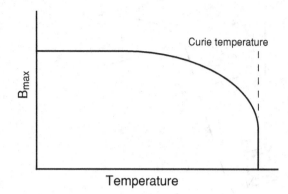

18.14. Decrease of saturation magnetization with temperature up to the Curie temperature.

so the direction of magnetization changes very little between neighboring atoms. There are two possibilities. In Bloch walls the direction of magnetization rotates in a plane parallel to the wall. Figure 18.15 illustrates a 180° domain wall and Figure 18.16 illustrates a 90° domain wall.

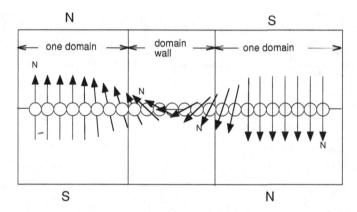

18.15. Schematic illustration of a 180° domain wall.

Magnetic oxides

Magnetite (loadstone) was the first known magnetic material. Its formula Fe_3O_4 may be written $Fe^{+2}Fe_2^{+3}O_4^{-2}$. Its structure is similar to spinel, $Mg^{+2}Al_2^{+3}O_4^{-2}$. In both of these structures oxygen ions are arranged in an fcc pattern. There are two types of sites for the anions: octahedral sites of sixfold coordination (Figure 18.17A) and tetrahedral sites of fourfold coordination (Figure 18.17B). The number of octahedral sites is the same as the number of oxygen ions and there are twice as many tetrahedral sites as oxygen ions. One of the two Fe^{+3} ions is in a tetrahedral site and the other is in an octahedral site. The Fe^{+2} ion is in an octahedral site. There are a number of similar compounds, known *as inverse spinels*,* where another divalent ion may substitute for Fe^{+2}. An example

---

* In a spinel (e.g., $MgAl_2O_4$) the M^{+2} ions are in tetrahedral sites and the M^{+3} ions are in octahedral sites.

MAGNETIC BEHAVIOR OF MATERIALS

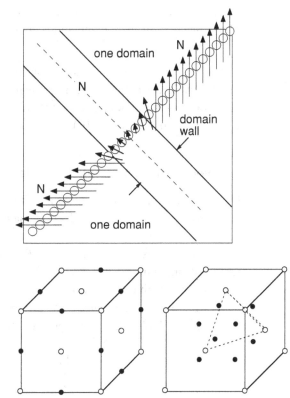

18.16. Schematic illustration of a 90° domain wall.

18.17. The black dots indicate the octahedral sites in A and the tetrahedral sites in B.

is $NiFe_2O_4$. The magnetic moment of the M^{+3} ion in the tetrahedral site is opposed to that of the M^{+3} ion in the octahedral site so the M^{+3} ions make no overall contribution to the magnetism. The sole contribution is from the M^{+2} ions in the tetrahedral sites. Table 18.3 lists the magnetic moments per atom in spinels.

Table 18.3. *Magnetic moments per atom in spinels*

Ion	3d electrons	Bohr magnetons
Fe^{+3}	$3d^5$	5
Mn^{+2}	$3d^5$	5
Fe^{+2}	$3d^6$	4
Co^{+2}	$3d^7$	3
Ni^{+2}	$3d^8$	2
Cu^{+2}	$3d^9$	1
Zn^{+3}	$3d^{10}$	0

Although zinc has no magnetic moment, it raises the magnetic moment of inverse spinels. Zn^{+3} ions occupy tetrahedral sites as they would in a normal spinel, thereby forcing Fe^{+3} ions into an octahedral site where they contribute to the magnetic moment.

EXAMPLE 18.2. Calculate the saturation magnetization of magnetite, $[Fe_3O_4]_8$. The lattice parameter of magnetite is about 0.839 nm.

SOLUTION: Only the Fe^{+2} ions contribute to the magnetization. There are 8 Fe^{+2} ions, each contributing 4 Bohr magnetons: $M = 3 \times 8/(0.839 \times 10^{-9}\text{m})^3$ (Bohr magnetons/m^3)(9.27×10^{-24})(Am^2/Bohr magneton) $= 3.77 \times 10^5$ A/m. $B_s = (4\pi \times 10^{-7})(3.77 \times 10^5) = 0.47$ T.

Soft versus hard magnetic materials

Most magnetic materials fall into one of two classes: soft and hard. Soft magnetic materials are easily magnetized and demagnetized. Hard magnetic materials are permanent magnets. They are difficult to magnetize and demagnetize. The hysteresis is very large. The remanence, B_r, and coercive force, H_c, are high. The terms *soft* and *hard* are historic. The best permanent magnets in the 1910s were made of martensitic steel, which is very hard, and the best soft magnets were made from pure annealed iron. The differences of the *B–H* curves are shown in Figure 18.18. Table 18.4 shows the extreme differences.

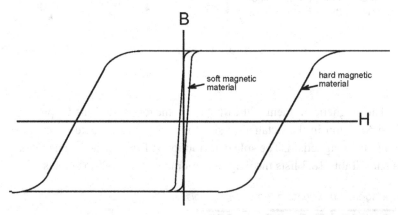

18.18. A hard magnetic material has a much greater hysteresis than a soft magnetic material. The differences are much greater than shown in this figure.

Soft magnetic materials

For a material to be soft magnetically, its domain walls must move easily. The principal obstacles to domain wall movement are inclusions and grain boundaries. Low dislocation contents, residual stresses, and a low interstitial content, are also important.

Inclusions are important obstacles to domain wall movement because the energy of the system is lowered more when a domain wall passes through an inclusion than when the boundary has separated from the inclusion. This is illustrated in Figure 18.19.

MAGNETIC BEHAVIOR OF MATERIALS

Table 18.4. Coercive forces of several materials

Material	Composition*	Coercive force
		H_c, A/m
Supermalloy	79% Ni, 5% Mo	.016
Oriented Si steel	3.25% Si	8.0
Hot-rolled Si steel	4.5% Si	40
Mild steel (normalized)	0.2% C	320
Carbon steel magnet	0.9% C, 1% Mn	4×10^3
Alnico V	24% Co, 14% Ni, 8% Al, 3% Cu	48×10^3
Alnico VIII	35% Co, 14.5% Ni, 7% Al, 5% Ti, 4.5% Cu	100×10^3
Barium ferrite	$BaO-6Fe_2O_3$	150×10^3
Bismanol	MnBi	290×10^3
Pt–Co	77% Pt, 23% Co	340×10^3

*Balance Fe.
Source: Data from J. K. Stanley, *Electrical and Magnetic Properties of Metals* (Materials Park, OH: ASM, 1943). Reprinted with permission of ASM International®. All rights reserved. www.asminternational.org.

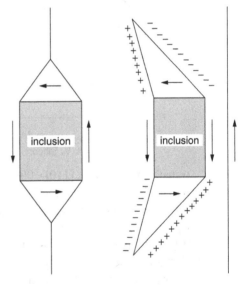

18.19. The difference between the domain boundaries at an inclusion depending on whether the inclusion lies on a boundary (left) or not (right). The total length of the boundary is lowered by the inclusion.

Uses of soft magnetic materials include transformers, motors and generator cores, solenoids, relays, magnetic shielding, and electromagnets for handling scrap. Many of these applications employ silicon iron (usually 3 to 3.5% Si). Alloys containing 3% or more silicon are ferritic at all temperatures up to the melting point. See Figure 18.20. Silicon increases the electrical resistance of iron. A high electrical resistance is desirable for transformers because eddy currents are one of the principal power losses in transformers. Remember power loss is inversely proportional to resistance ($P = EI = E^2/R$). Use of thin sheets also minimizes eddy current losses.

It is possible to control the crystallographic texture of silicon iron sheet by controlling the rolling and heat-treating schedules. The usual texture for the

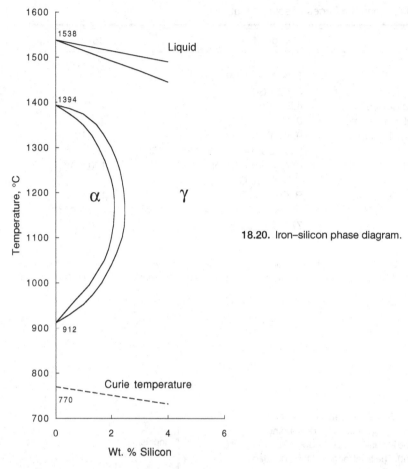

18.20. Iron–silicon phase diagram.

transformer sheets is {110}<001>, which is called the *Goss* texture. This texture has the <001> easy direction of magnetization aligned with the prior rolling direction. Transformers can be made so that they will be magnetized in a <001> direction. The cube texture, {100}<001>, is even more desirable, but it is more difficult to produce. Both are illustrated schematically in Figure 18.21.

Core losses decrease with increasing silicon content and increase with increasing frequency. Oxide materials are very useful at high frequency because power losses decrease with increasing electrical resistance.

For very soft magnetic magnets, magnetostriction should be minimized. The reason is that magnetostriction causes dimensional incompatibilities at 90° domain boundaries that must be accommodated by elastic straining of the lattice. This is illustrated in Figure 18.22. In iron–nickel alloys the magnetostriction and the magnetocrystalline anisotropy are very low at about 78% Ni. Iron–nickel alloys have very high initial permeabilities. Mu metal (75% Ni), permalloy (79% Ni), and supermalloy (79% Ni, 4% Mo) are examples that find use in audio transformers.

A metallic glass containing 80% Fe and 20% B is an excellent soft magnetic material because there are no grain boundaries to obstruct domain wall motion.

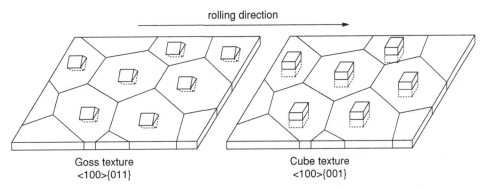

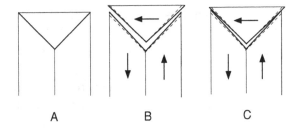

18.21. Textures in silicon. In both the Goss and cube textures the <100> direction is aligned with the rolling direction. The {011} is parallel to the sheet in the Goss texture and the {001} is parallel to the sheet in the cube texture.

18.22. In iron, magnetostriction causes an elongation in the direction of magnetization. This creates a misfit along 90° domain boundaries, which must be accommodated elastically.

Hard magnetic materials

A high H_c coercive force is desirable for hard magnets, but most important is a high $H \times B$ product. The second quadrant of the B–H curve (Figure 18.23) is most important. Often the maximum $B \times H$ product (Figure 18.24) is taken as a figure of merit.

18.23. Second quadrant of B–H curves for selected alloys. From R. M. Rose, L. A. Shepard, and J. Wulff, *The Structure and Properties of Materials*, vol. IV, *Electronic Properties* (New York: Wiley, 1996). Reprinted with permission of John Wiley & Sons, Inc.

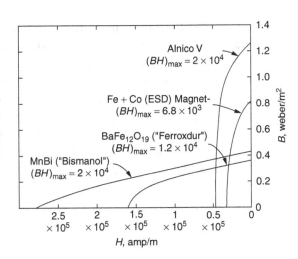

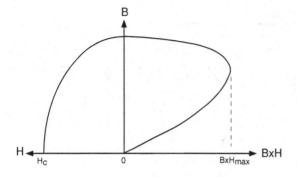

18.24. Second quadrant of a B–H curve (left) and the corresponding B × H product (right). The maximum B × H product is a figure of merit.

High $B \times H$ products are promoted by

1. small, isolated particles that are single domains.
2. elongated particles
3. a high magneto-crystalline energy

In a microstructure consisting of small isolated particles surrounded by a non-ferromagnetic phase, there are no domain walls that can move. The direction of magnetization can be changed only by rotating the magnetization out of the easy direction into another equivalent easy direction. If there is a high magnetocrystalline energy, this will require a high field. Hexagonal structures are useful here because there are only two easy directions, [0001] and [000$\bar{1}$], which differ by 180°. When ferromagnetic particles are elongated, the intermediate stage will have a high magnetostatic energy. Figure 18.25 illustrates this.

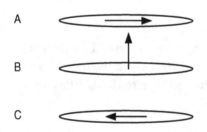

18.25. As the direction of magnetization of an elongated particle is reversed (from A to C), it must be magnetized in a direction that increases its magnetostatic and magnetocrystalline energies (B).

Among the most popular magnet materials are aluminum–nickel–cobalt–iron alloys called alnico. Alnico V contains 8% Al, 14.5% Ni, 23% Co, 3% Cu, and 0.5% Ti with the balance Fe. At very high temperatures it is a single bcc phase but it decomposes into two bcc phases below 800 °C. The phase high in Co and Fe is ferromagnetic and it precipitates as fine particles. If the precipitation occurs in a magnetic field, the particles are elongated (Figure 18.26A) whereas they are equiaxed in the absence of a field (Figure 18.26B). The difference in the B–H curves is shown in Figure 18.27.

Some of the best hard magnetic materials are those with a hexagonal structure. In these there are only two possible domains, differing by 180°. Table 18.5 lists the maximum BH product for several alloys. Cheap permanent magnets can be made by aligning fine iron powder in a magnetic field while it is being bonded by rubber or a polymer.

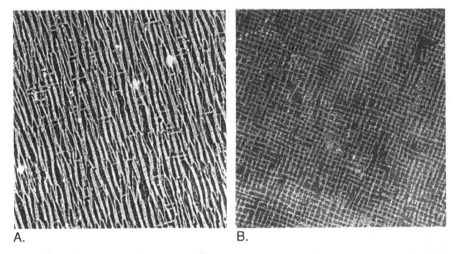

18.26. Microstructure of Alnico V (~50,000X) after precipitation in a magnetic field (left) and in the absence of a magnetic field (right). From R. M. Rose, L. A. Shepard, and J. Wulff, *Structure and Properties of Materials,* vol. IV, *Electronic Properties* (New York: Wiley, 1966). Reprinted with permission of John Wiley & Sons, Inc.

18.27. The effect of heat treating in a magnetic field on the demagnetization curves for Alnico V.

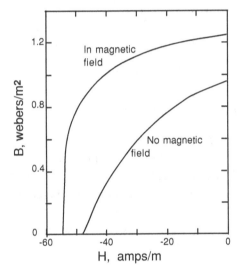

Square-loop materials

Magnetic materials used for memory storage are neither hard nor extremely soft. They must be soft enough to have their direction of magnetization changed by

Table 18.5. Maximum *BH* products for several alloys

Samarium–cobalt	120,000 A – Wb/m^3
Platinum–cobalt	70,000
Alnico	36,000
Carbon steel	1,500

currents in computer circuits but hard enough to be unaffected by stray magnetic fields. Square *B–H* curves are desirable. By heat treating permalloy and supermalloy in a magnetic field, a texture can be formed which has a square-loop hysteresis curve (Figure 18.28). This is useful in logic circuits where the material is magnetized in one direction or the other. For high fidelity transformers, linearity (constant μ) is needed.

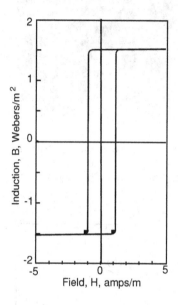

18.28. A square-loop hysteresis curve for permalloy can be obtained by heat treating in a magnetic field.

NOTES OF INTEREST

1. The word *magnetite* comes from Magnesia, which is the region in Asia Minor from which magnetite first came.
2. The specific heats of ferromagnetic materials show an anomalous specific heat near their Curie temperatures. Figure 18.29 shows a spike of the specific heat of

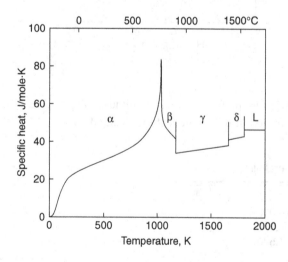

18.29. The specific heat of iron shows a peak near the Curie temperature. From W. F. Hosford, *Physical Metallurgy* (Boca Raton, FL: CRC Press, 2005), p. 331, figure 15.2.

iron near its Curie temperature. When this was first observed, it was mistaken for a latent heat of transformation to a new phase, β, in temperatures between the α and γ phases.
3. A magnet provides a simple way of distinguishing austenitic stainless steels from the other grades.

REFERENCES

E. A. Nesbitt. *Ferromagnetic Domains*. Murray Hill, NJ: Bell Telephone Laboratories, 1962.

R. M. Rose, L. A. Shepard, and J. Wulff. *The Structure and Properties of Materials*, vol. IV, *Electronic Properties*. New York: Wiley, 1966.

J. K. Stanley. *Electrical and Magnetic Properties of Metals*. Materials Park, OH: ASM, 1963.

PROBLEMS

1. Metallic nickel has a saturation magnetization of 0.6 Wb/m^2 and a lattice parameter of 0.352 nm. What is the magnetic moment per atom in Bohr magnetons?

2. Find the $(B \times H)_{max}$ product for the Alnico V heat treated in a magnetic field from Figure 18.27.

3. How many different domain orientations are possible in (A) iron, (B) nickel, and (C) cobalt?

4. Explain in terms of your answer to Problem 3 why many of the hard magnetic alloys are based on cobalt.

5. The density of iron is 7.87 Mg/m^3 and its saturation magnetization at 0 K is 2.16 T. Find the number of Bohr magnetons per atom.

6. The Goss texture in silicon iron sheets has a <110> direction aligned with the prior rolling direction and a {001} plane aligned with the plane of the sheet. In the cube texture a <100> direction is aligned with the prior rolling direction and a {001} plane is aligned with the plane of the sheet. For iron, the permeability, μ, is highest in the <100> direction. Plot schematically how μ varies with the angle, θ, from the rolling direction for both the cube and Goss textures. Extend your plot from 0 to 180°.

7. Why are magnetic oxides preferred for the cores of very high frequency transformers?

19 Porous and Novel Materials

Applications of porous materials

There are many applications of porous materials. Their ability to fill space with a minimum weight leads to their use in packaging. Life jackets and rafts use porous materials because of their low density. Examples of their use as thermal insulators range from Styrofoam cups to heat shields for space craft. Aluminum baseball bats are filled with foam to dampen vibrations. The low elastic moduli and high elastic strains of foams lead to use as cushions and mattresses. Filters are made from porous materials.

Stiff lightweight structures such as aircraft wings are made from sandwiches of continuous sheets filled with foams or honeycombs. Open porous structures can form frameworks for infiltration by other materials leading to application of biocompatible implants. Open pore structures are used as supports for catalysts.

Fabrication of porous foams

Natural cellular materials include sponges and wood. Foams of polymers, metals, and ceramics can be made by numerous methods. Foams are often produced by entrapping evolved gas. Inert gasses such as CO_2 and N_2 may be dissolved under high pressure and released by decreasing the pressure. Gas bubbles may also be formed by chemical decomposition or chemical reaction. Mechanical beating will produce foams. Foamed structures may be formed by bonding previously expanded spheres as in the case of polystyrene. Incomplete sintering of pressed powders creates materials with continuous internal passages that find use as filters and oilless bearings.

Metallic and ceramic foams are often made from polymeric precursors. Ceramic foams can be made by dipping a polymer foam into a ceramic slurry, drying, and then sintering at a high enough temperature to decompose the polymer. Metallic foams can be made by electroless plating of metal on a polymer foam and subsequent heating to drive off the polymer. Finally, carbonaceous foams can be made by pyrolyzing polymeric foams.

Spinodal decomposition usually results in two continuous phases. If one of these is etched away, the result is a porous structure. Filters of glass are produced this way.

Morphology of foams

There are two types of foams: closed cell foams and open cell (or reticulated) foams. In open foams, air or other fluids are free to circulate. These are used for filters and as skeletons. They are often made by collapsing the walls of closed cell foams. Closed cell foams are much stiffer and stronger than open cell foams because compression is partially resisted by increased air pressure inside the cells. Figure 19.1 shows that the geometry of open and closed cell foams can modeled by Kelvin tetrakaidecahedra.

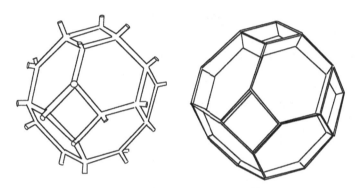

19.1. Open and closed cell foams modeled by tetrakaidecahedra.

Relative density of foams

The parameter ρ^*/ρ_s, where ρ^* is the density of the cellular structure and ρ_s is the density of the solid material from which it is made, is called the *relative density*. Equations 19.1 and 19.2 are first approximations for relative density if the thicknesses of the walls and ligaments, t, is much smaller than the cell length, l, ($t << l$).

For open cell foams

$$\rho^*/\rho_s = C_2(t/l)^2, \tag{19.1}$$

and for closed cell foams

$$\rho^*/\rho_s = C_3(t/l), \tag{19.2}$$

where C_2 and C_3 are constants.

However, these equations need corrections for double counting at edges and corners. A reasonable approximation for all of the structures is given by

$$\rho^*/\rho_s = 1.2[t_e/l]^2 + 0.7(t_f/l), \tag{19.3}$$

where t_e and t_f are the thicknesses of the edges and faces.

Structural mechanical properties

The elastic stiffness depends on the relative density. In general the dependence of relative stiffness, E^*/E, where E^* is the elastic modulus of the structure and E is the modulus of the solid material on relative density, is of the form

$$E^*/E = (\rho^*/\rho_s)^n. \tag{19.4}$$

Experimental results shown in Figure 19.2 indicate that for open cells $n = 2$ so

$$E^*/E = (\rho^*/\rho_s)^2. \tag{19.5}$$

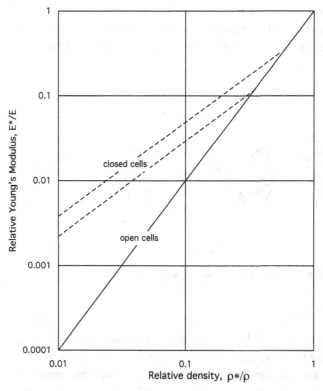

19.2. Data from L. J. Gibson and M. F. Ashby. *Cellular Foams* (Cambridge, U.K.: Cambridge Univ. Press, 1999).

For closed cell foams, E^*/E is much higher and $n < 2$. While deformation under compression of open cell foams is primarily by ligament bending, compression of closed cell wall foams involves gas compression and wall stretching in addition to wall bending as well.

Honeycombs

Honeycombs can be made by folding and gluing thin sheets, gluing and expanding thin sheets, casting, and extruding. Figure 19.3 illustrates a honeycomb structure. Panels with a high bending stiffness-to-weight ratio are often made with a honeycomb structure sandwiched between two sheets or plates. While the honeycomb does not directly contribute much to the stiffness, it separates the outer sheets so they have a maximum bending resistance.

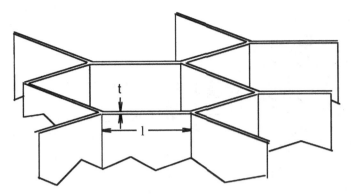

19.3. Hexagonal honeycomb structure.

Novel structures

A polyester foam with a negative Poisson's ratio has been reported by Lakes.[*] Its cells consist of reentrant faces. Figure 19.4 shows a cell with 24 faces. When the cell is extended its sides are moved inward and when it is compressed the walls move outward. The negativity of ν increases with θ.

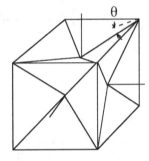

19.4. A cell structure with a negative Poisson's ratio. If it is compressed, the sides will move inward.

Sigmund and Torquato[†] devised two-dimensional composite materials with negative coefficients of thermal expansion by combining a material with a high coefficient of thermal expansion with one having a low coefficient of thermal expansion. Figure 19.5 shows an example of such a composite designed by Chen et al.[‡] As the material is heated, the high thermal expansion causes the horizontal and vertical bars to move inward.

NOTES OF INTEREST

1. J. Qui and J. W. Halloran[**] produced the cross section in Figure 19.6 by extruding a composite of oxides of Fe-36 atom % Ni($\alpha = 3 \times 10^6/^\circ$C) and Fe-60 atom % Ni($\alpha = 14 \times 10^{-6}/^\circ$C) and carbon (for the void). Firing reduced the oxides and burned out the carbon. The result was a composite with a

---

[*] R. S. Lake, *Science* 235 (1987): xxx.
[†] O. Sigmund and S. Torquato, *J. Mech. Phys Solids*, 45 (1997): 1037–67.
[‡] B.-C. Chen, E. C. N. Silva, and N. Kikuchi, *Int. J. Numer. Meth. Eng.* 52 (2001): 23.
[**] J. Qui and J. W. Halloran, *J. Mater. Sci.* 39 (2004): A113–4118.

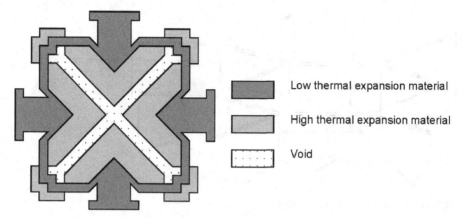

19.5. Composite material with a negative coefficient of thermal expansion.

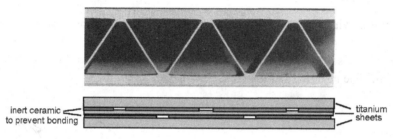

19.6. A wing panel made of diffusion-bonded and superplastically formed titanium.

coefficient of thermal expansion of $\alpha = -3 \times 10^{-6}/°C$ compared to the prediction of $\alpha = -3.2 \times 10^{-6}/°C$.

2. Diffusion bonding combined with superplastic forming provides a novel alternative to sandwiches filled with a honeycomb for light structures with high bending resistance. Figure 19.6 shows the cross section of such a structure. It was made by diffusion bonding three sheets of titanium. Bonding was prevented in certain areas by coating with an inert ceramic coating. The structure was then expanded by forcing an inert gas into the unbonded channels. Superplastic behavior was required because of the severe elongations of the interior ligaments.

REFERENCE

L. J. Gibson and M. F. Ashby. *Cellular Foams*. Cambridge, U.K.: Cambrige Univ. Press, 1999.

PROBLEMS

1. Calculate ν for the structure in Figure 19.4 if $\theta = 15°$, assuming that the diagonal bars are very stiff compared to the bars on the cube edges and that the dimensions are as shown in Figure 19.7.

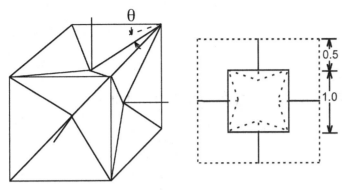

19.7. Unit cell for Problem 1.

2. The relative density of honeycombs is given by $\rho^*/\rho_s = C_1(t/l)$. Determine the value of C_1 for the hexagonal honeycomb in Figure 19.6. Neglect the tops and bottoms of the cells.

20 Shape Memory and Superelasticity

Shape memory alloys

With the shape memory effect, heating reverses prior plastic deformation. Alloys that exhibit this effect are ordered solid solutions that undergo a martensitic transformation on cooling. Shape memory effects were first observed in AuCd in 1932, but it was the discovery by Buehler et al. in of the effect in NiTi in 1962 that stimulated interest in shape memory. The alloy TiNi (49 to 51 atomic % Ni) has an ordered bcc structure at 200 °C. On cooling it transforms to a monoclinic structure by a martensitic shear. The shear strain associated with this transformation is about 12%. There is more than one variant of the transformation. If only one variant of the martensite were formed, the strain in the neighboring untransformed lattice would be far too high to accommodate. Instead two mirror image variants form in such a way that there is no macroscopic strain. The macroscopic shape is the same as before the transformation. Figure 20.1 illustrates this. The boundaries between the two variants are highly mobile. The resulting structure can deform easily by movement of these boundaries. Figure 20.2 shows stress–strain curves above and below the M_s. Heating the deformed material above the A_f temperature causes it to transform back to the ordered cubic structure by martensitic shear. The overall effect is that the deformation imposed on the low temperature martensitic form is reversed on heating. The critical temperatures for reversal in TiNi alloys are typically in the range of 80 to 100 °C but are sensitive to very minor changes in composition so material can be produced with specific reversal temperatures. Excess nickel greatly lowers the transformation temperature. It is also depressed by small additions of iron and chromium. Copper decreases the hysteresis.

Other shape memory alloys are listed in Table 20.1, and Figure 20.3 shows the dependence of the M_s temperature for Cu–Zn–Al alloys on composition. For the copper-base alloys controlled cooling is necessary after heating into the β phase region.

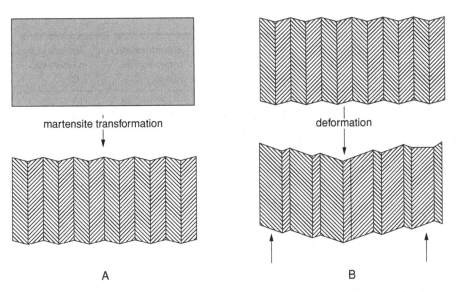

20.1. (A) As the material is cooled it undergoes a martensitic transformation. By transforming to equal amounts of two variants, the macroscopic shape is retained. (B) Deformation occurs by movement of variant boundaries so the more favorably oriented variant grows at the expense of the other. Reprinted with permission of Cambridge University Press from W. F. Hosford, *Mechanical Behavior of Materials* (New York: Cambridge Univ. Press, 2005).

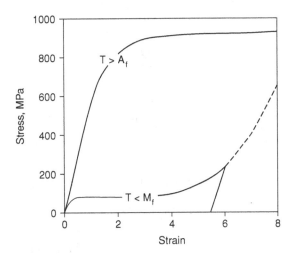

20.2. Stress–strain curve for a shape memory material. The lower curve is for deformation when the material is entirely martensitic. The deformation occurs by movement of variant boundaries. After all of the material is of one variant, the stress rises rapidly. The upper curve is for the material above its A_f temperature. Adapted from a sketch by D. Grummon.

Superelasticity

This phenomenon is closely related to the shape memory. Applied stress raises the A_f, A_s, M_s, and M_f temperatures, as illustrated in Figure 20.4, so deformation at temperatures slightly above the A_f will cause the material to transform by martensitic shear to its low temperature form. Once the stress is released, the material will revert to the high temperature form by reversing the martensitic shear. A stress–strain curve for Fe_3Be is shown in Figure 20.5.

Table 20.1. Alloys with shape memory

Alloy	Composition	Transformation temperature range, °C	Temperature hysteresis, °C
AuCd	46.5 to 50 at. %Cd	30 to 100	15
Cu–Al–Ni	14 to 14.5 wt. %Al 3 to 4.5% wt. %Ni	−140 to 100	35
Cu–Zn–x x = a few wt. % Si, Sn or Al.	38.5 to 41.5% wt. Zn	−180 to 200	10
In–Ti	18–23% at. %Ti	60 to 100	4
Ni–Al	36 to 38% at. %Al	−180 to 100	10
NiTi	49–51 at. %Ni	−50 to 110	30

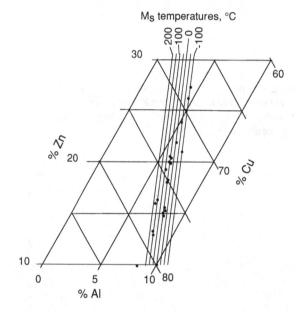

20.3. The dependence of the M_s temperature of copper–zinc–aluminum alloys on composition. The dots indicate alloys for which the M_s has been measured. Data from D. E. Hodgson, M. H. Wu, and R. J. Biermann, *Shape Memory Alloys*, Johnson Matthey, http://www.jmmedical.com/html/_shape_memory_alloys_html (accessed May 6, 2006).

According to the Clausius–Clapeyron equation,

$$d(A_f)/d\sigma = T\varepsilon_o/\Delta H, \qquad (20.1)$$

where ε_o is the normal strain associated with the transformation and ΔH is the latent heat of transformation (about 20 J/g for TiNi). Note that the terms $d(A_s)/d\sigma$, $d(M_s)/d\sigma$, and $d(M_f)/d\sigma$ could be substituted for $d(A_f)/d\sigma$ in Equation 20.1.

Figures 20.6 and 20.7 illustrate the relation between shape memory and superelasticity. Shape memory occurs when the deformation takes place at a temperature below the M_f. Superelasticity occurs when the deformation is at a temperature above the A_f. For both the memory effect and superelasticity, the alloy must be ordered, there must be a martensitic transformation, and the variant boundaries must be mobile.

20.4. As stress is applied to a superelastic material, the A_f, A_s, M_s, and M_f temperatures for the material all increase so that the material undergoes martensitic shear strains. When the stress is removed, the material reverts to its high temperature form, reversing all of the martensitic deformation. Adapted from a sketch by D. Grummon.

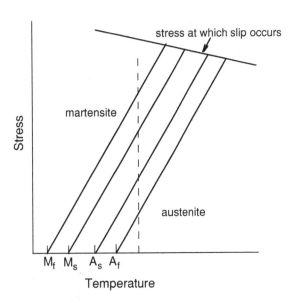

20.5. The stress–strain curve for superelastic Fe_3Be. After the initial Hookean strain, the material deforms by martensitic transformation. On unloading the reverse martensitic transformation occurs at a lower stress. Adapted from R. H. Richman, in *Deformation Twinning* (New York: AIME, 1963), p. 267, figure 23.

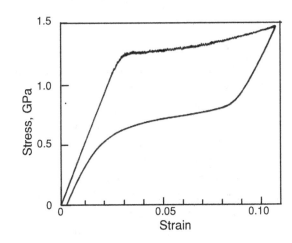

20.6. Schematic stress–strain curves. Superelasticity occurs at temperatures above the A_f. The deformation in shape memory occurs below the M_f temperature. Reversion requires heating above the A_f.

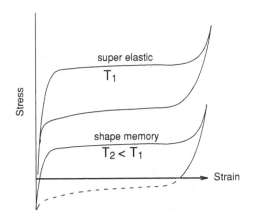

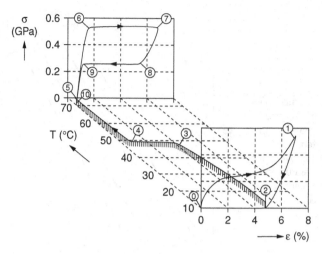

20.7. Schematic illustration of the difference between shape memory and superelastic effects. For shape memory, the deformation occurs at a temperature for which the material is martensitic. A superelastic effect occurs when the deformation occurs just above the A_f temperature. From J. A. Shaw, *Int. J. Plasticity* 16 (2000): 542.

Applications

Superplastic applications of TiNi include stents for keeping arteries open and couplings for sealing tubes or pipes. The very high power-to-weight ratio of the transformation may lead to other applications. On–off actuators can be constructed by combining two materials with different transition temperatures. These find uses in temperature control systems. Single shape memory materials may be "trained" to remember two shapes: one at the high temperature and one at the low temperature. This is possible because of local stress set up by small untransformed regions. Unfortunately, the shape memory tends to degrade after repeated cycles and cycling can lead to fatigue failure.

Eyeglass frames made from superelastic NiTi absorb very large deformations without damage. Medical uses of NiTi include guide wires for steering catheters in the body and wires for orthodontic corrections.

Shape memory in polymers

When a polymer is deformed at temperatures below its glass transition and then heated above the glass transition temperature, its shape will revert to that it originally had before being deformed. The amount of reversible strain is much larger than in metals (up to 400% for polymers vs. less than 10% for metals). This effect is utilized in shrink-wrapping of consumer products. Films are stretched biaxially at temperatures below their glass transition temperature. After the product is wrapped, the temperature is raised above the glass transition temperature by warm air, allowing the film to shrink tightly around the product. This effect is also used for insulating wiring joints. Preexpanded tubes are slipped over the joints. Heating causes them to shrink tightly around the bare wires.

Medical applications of biodegradable shape memory polymers include their use for removing blood clots formed during strokes. Preshaped foams can be used to fill cranial aneurisms. Loosely tied sutures made from fibers that have been stretched at 50 °C will tighten when heated just above room temperature.

NOTE OF INTEREST

The discovery of the shape memory effect in TiNi by Buehler et al. at the Naval Ordinance Labs occurred during an investigation of the alloy for possible use as a corrosion-resistant knife for underwater activities. The investigators called the alloy *nitinol* for Nickel, Titanium, and Naval Ordinance Labs.

REFERENCES

W. J. Buehler, J. V. Gilfrich, and R. C. Wiley. *J. Appl. Phys.* 34 (1963): 1475–7.
J. D. Harrison and D. E. Hodgson. *Shape Memory Effects in Alloys*. New York: Plenum Press, 1975.
Materials for Smart Systems. Pittsburgh, PA: Materials Research Society, 1995.
J. A. Shaw. *Int. J. Plasticity* 16 (2000): 451–62.
K. Shimizu and T. Tadaki. *Shape Memory Alloys*. New York: Gordon and Breach, 1987.

PROBLEMS

1. Estimate the stress required to shift the A_s temperature of a NiTi alloy from 100 to 80 °C.

2. Estimate the amount of elastic energy per volume that can be stored at 75 °C in the material shown in Figure 20.7.

3. Figure 20.3 shows that increased amounts of both aluminum and zinc lower the M_s temperatures of copper–zinc–aluminum alloys. For both aluminum and zinc, determine dT/dc where T is the M_s temperature and c is the atomic % solute. Note that the compositions in Figure 20.3 are in wt. %.

21 Calculations

This chapter is intended to help engineering students solve engineering problems. A suggested procedure is:

1. Estimate the answer before you start. This will provide you with something to check your answer against.
2. Consider making a sketch. Very often this will clarify the problem.
3. Define variables and select an appropriate form of analysis. If a numerical answer is required, this may involve developing an algorithm or selecting appropriate equations.
4. Be sure to include units.
5. Do the algebra before substituting numbers. Often things drop out. This makes numerical calculations simpler.
6. Find the solution and check it against the original estimate.
7. Report your answer with an appropriate number of significant figures.

Estimates

When attempting to solve an engineering problem first it is helpful to make an estimate of the final answer. An initial estimate provides a check to final answers.

Rough estimates can be made from human experience. One knows that most solids sink when immersed in water so they have densities greater than 1 Mg/m^3. Also almost all solids have densities less than 20 Mg/m^3. Those numbers form reasonable bounds for the density of most materials. Of the solids that float in water, most float with less than half of the solid above water. That means they have densities between 1 and 0.5 Mg/m^3. Most plastics have densities over 0.9 Mg/m^3.

Estimation is necessary for making reasonable assumptions. For example, during heat treatment of a metal, grain growth may occur. This growth releases energy, which will go into heat. A simple calculation will let one know whether it is reasonable to neglect this when calculating the power needed to run the furnace.

Often a crude estimation will be enough to decide which of several alternatives should be considered. For example, if an item is to be made by casting, several different casting processes are possible. An estimation of the total production is necessary to decide whether die casting, sand casting, or permanent mold casting is best. Die casting is the fastest and most automated but requires the greatest capital investment. It is appropriate only for very high quantities. Sand casting is the slowest and involves the most labor. It is appropriate for making a few parts. Permanent mold casting is intermediate and may be best for intermediate quantities.

EXAMPLE 21.1. Estimate the maximum load that a passenger elevator must be designed to carry. This requires estimating the maximum weight of passengers and the number of passengers. It would be reasonable to assume 250 lbs per person and that a 250-lb person would occupy 1.5 ft^2. If the elevator measures 5×6 ft $= 30$ ft^2, 20 people might squeeze in so the load would be 20 persons $\times$ 250 lb/person $=$ 5000 lb. Let's not quibble about the estimates of 20 persons and 250 lb/person. This is only a rough estimate. Similarly, the necessary load-carrying capacity of a bridge can be estimated by multiplying the maximum number of vehicles that could fit on the bridge multiplied by the average weight per vehicle. In both cases a factor of safety can be used to cover uncertainties in our estimates.

EXAMPLE 21.2. Estimate the annual U.S. production of soup cans. Here the guesses are less certain. We might assume that the average person has soup once every two weeks. With a U.S. population of 300 million the consumption would be $(300 \times 10^6$ persons$)(365$ days per year$)(1$ can per person/14 days$) = 8$ billion cans/year. If we needed a more accurate number, we could use the Web to find a source that might have figures from can makers.

Estimates are needed to make experiments. If we are to measure an electrical current, a pressure, a weight, or a length we must first make a crude estimate of the answer to determine what equipment to use. Ammeters, pressure gauges, and weighing devices all come in different sizes. An analytical balance that could weigh an empty can cannot weigh an automobile. Odometers, yardsticks, and micrometers all measure length but are not interchangeable.

Sketches

In attacking many problems, the very first thing that should be done is to make a sketch. Pictorial visualization is helpful and often necessary if the problem involves geometry. Many of the greatest engineers were excellent draftsmen. Leonardo da Vinci and Michelangelo were the best engineers of their period. The original drawings of the Brooklyn Bridge by the chief engineer, Roebling, are now being preserved in an art museum in Brooklyn. Feynman diagrams, devised by Nobel physicist Richard Feynman to represent nuclear reactions, conceptually are one of his greatest contributions to physics. Engineers need not be artists. They need only to be able to make freehand sketches that they themselves can interpret correctly. Although it is better if others can also understand the drawings, it is essential only that the drawer can.

Sketches are important in dealing with a wide variety of problems including electrical circuits, crystal structures, force and moment balances, chemical processes, and flow of air around airfoil or computer programs.

EXAMPLE 21.3. If one is concerned with blanking 6-in.-diameter circles from a wide sheet, a sketch would help determine the width of a wide sheet that would minimize the amount of scrap per blank (Figure 21.1).

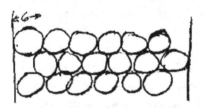

21.1. Sketch of blank layout.

EXAMPLE 21.4. The contact length between a roll and a work piece can be visualized with a sketch (Figure 21.2).

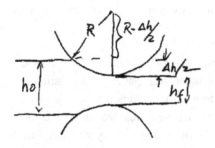

21.2. Sketch of deformation zone in rolling.

EXAMPLE 21.5. Still another example is finding the resistance of a network of resistors (Figure 21.3).

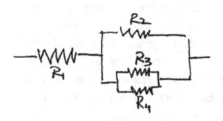

21.3. Arrangement of resistors.

EXAMPLE 21.6. A sketch of a force balance shows the relation between pressure, diameter, and wall thickness of a pressurized sphere Figure 21.4 shows that $P(\pi d^2/4) = \pi D t \sigma_x$.

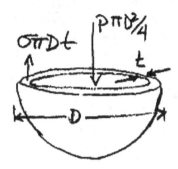

21.4. Force balance on half a sphere.

EXAMPLE 21.7. A student cannot remember whether the relation of the lattice parameter of a bcc crystal to the atomic radius is $a = 4r/\sqrt{3}$ or $a = \sqrt{3}r/4$. A simple sketch (Figure 21.5) would show that $a = 4r/\sqrt{3}$.

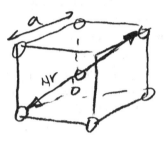

21.5. Sketch of body-centered cubic unit cell.

Units

Units are a necessary part of all numerical answers. To express the water consumption of a process as 23 is meaningless. Does that mean 23 gal, 23 m³, 23 ft³, 23 L, or something else? A salary of $100 could mean $100/week, $100/d or $100/h. One should never assume that the reader will know the correct units.

Every equation must be dimensionally correct. The dimensions on the right-hand side must be the same as those on the left-hand side. Suppose we know that the conductivity, σ, of a material may be expressed as $\sigma = nq\mu$, where n is the number of carriers per volume, q is the charge per carrier (coulombs/carrier), and μ is the mobility, but we do not know the units of mobility. If conductivity, σ, is expressed in (ohm · m)$^{-1}$, the units of mobility must be $[\sigma(\text{ohm} \cdot \text{m})^{-1}]/[(n \text{ carriers/m}^3)(q \text{ C/carrier})] = \mu \text{ m}^2/(\text{V/s})$. This in turn can be thought of as a drift velocity, (m/s), divided by the voltage gradient, (V/m).

The values of mathematical functions are dimensionless. Quantities such as $\sin(\theta)$, $\exp(x^2/Dt)$, $\arctan(x/L)$, $\ln(\dot{\varepsilon}/\dot{\varepsilon}_0)$, and $\ln(x/L)$ are dimensionless. Exponents and the arguments of functions must also be dimensionless. If x were to have the units of length, y^x, $\sin(x)$, and $\ln(x)$ would all have crazy units. Sometimes the units are hidden in constants. The strain–rate dependence of strength is sometimes expressed as $\sigma = \sigma_o + b \ln(\dot{\varepsilon})$. Since $\dot{\varepsilon}$ has the dimensions of 1/time, it would appear that the units of b must be (stress)[ln(time)]. However, this equation is more properly written as $\sigma = \sigma_o + b \ln(\dot{\varepsilon}/\dot{\varepsilon}_o)$, where $\dot{\varepsilon}_o$ is also a constant. Use of the first form implies that $\dot{\varepsilon}_o = 1/s$ and $\dot{\varepsilon}$ is expressed in s^{-1}.

The term $\operatorname{erf}[x/(2\sqrt{Dt})]$ often arises in problems involving diffusion. Here erf is called the error function. Note the quantity $x/\sqrt{Dt}$ is dimensionless if D is expressed in m^2/s, t in s, and x in m. Similarly, in the equation expressing the equilibrium number of vacancies per volume, $N_v = N_o \exp[-Q/(RT)]$, the term $Q/(RT)$ must be dimensionless. It is if Q is expressed in J/mol, R in J/(mol·K), and T in K. The equation $L = \tan(D/L)$ is dimensionally incorrect because the right-hand side is dimensionless but left-hand side has units of length.

Often it is useful to combine variables that affect physical phenomenon into dimensionless parameters. For example, the transition from laminar to turbulent flow in a pipe depends on the Reynolds number, $R_e = \rho L v/\mu$, where ρ is the fluid density, L is a characteristic dimension of the pipe, v is the velocity of flow, and μ is the viscosity of the fluid. Experiments show that the transition from laminar to turbulent flow occurs at the same value of R_e for different fluids, flow velocities, and pipe sizes. Analyzing dimensions is made easier if we designate mass as M, length as L, time as t, and force as F. With this notation, the dimensions of the variables in R_e are ML^{-3} for ρ, (L) for L, (L/t) for v, and $(FL^{-2}t)$ for μ. Combining these it is apparent that $R_e = \rho L v/\mu$ is dimensionless.

In problems involving non-steady-state heat conduction, the temperature distribution depends on a parameter $x/[Kt/(\rho C)]$, where x is the distance (L) from a location, t is the time (t), K is the thermal conductivity with dimensions of energy/[(time)(area)(temperature gradient)]$(ET^{-1}L^{-1}t^{-1})$, where E is energy. Here C is the heat capacity with units of $EM^{-1}T^{-1}$, ρ is density, ML^{-3}, x is distance (L), and t is time (t). For the same boundary conditions the same temperature will be found at the same times and locations, with different materials, if $x/[Kt/(\rho C)]$ has the same value.

Often one can solve problems simply by keeping track of units. The density of copper can be calculated as

density = mass/volume = (mass/unit cell)/(volume/unit cell)

mass/unit cell = (4 Cu atoms/uc)(643g/mol)/(6.023 × 10^{23} Cu atom/mol.)

Note this is

g/uc

volume/unit cell = $(0.3615 \times 10^{-9})^3 m^3/uc$

so the density is

[(4 Cu atoms/uc)(643.g/mol)/(6.023 × 10^{23} Cu atom/mol)]/
[$0.3615 \times 10^{-9})^3 m^3/uc$] = 8.94×10^6 g/m^3 or 8.94 Mg/m^3.

Note that if we cannot remember whether to multiply or divide by Avrogadro's number, the units will tell us.

Expressing quantities in detailed units is often of key importance.

EXAMPLE 21.8. Suppose we have a composite material containing two different materials, A and B. The weight percentages are 14% A and 86% B. Material A contains 23 wt% Fe and material B contains 12 wt% Fe. To find the total amount of iron in a gram of composite,

$$(0.14 \text{ g of A/g comp})(0.23 \text{ g Fe/g A}) + (0.86 \text{ g B/g comp})(0.12 \text{gFe/g B})$$
$$= 0.135 \text{ g Fe/gm composite.}$$

Note that by writing 0.23 g Fe/g A instead of just 23% the units work out completely.

In converting Celsius to Kelvin one should realize that an *interval* of one degree on the Celsius scale is exactly the same as an *interval* of one degree on the Kelvin scale. The linear coefficient of thermal expansion of iron may be written as 11.76×10^{-6}(m/m)/K or as 11.76×10^{-6}(m/m)/°C. One may think of this as (ΔT)K $= (\Delta T)$°C. No conversion is necessary. However, when dealing with absolute temperatures as in the perfect gas law, $PV = nRT$, or an Arrhenius rate equation, rate $= A \exp[-Q/(RT)]$, the temperature must be in Kelvin. The conversion $T(\text{K}) = T(°\text{C}) + 273$ is necessary.

If a fractional change of a linear dimension, $\Delta L/L$, is small and the same change occurs in all directions, then $\Delta V/V = 3\Delta L/L$. Thus, for iron, the thermal coefficient of volume expansion is $3 \times 11.76 \times 10^{-6}$(m/m)/°C $= 35.3 \times 10^{-6}$(m^3/m^3)/°C.

Available data

When problems are assigned in the classroom, the student is very often given all of the data necessary to solve the problem and no extra data. The real world is full of data, most of it irrelevant to the problem. The engineer must decide which data is appropriate to his/her own problem.

If there seems to be too little data, the first question is whether the missing data are really essential to solving the problem. If the answer is yes, then a search should be made through appropriate sources (handbooks, texts, etc).

If the missing data are essential and cannot be found, then there are still two possibilities. The problem can be solved in terms of the missing value by assigning a symbol to it. For example, the solution might be $37.2Y$ where Y is the yield strength of the alloy in MPa. The other possibility is to estimate the missing value and assume it in the calculation. In this case the assumption should be made clear to the reader.

Sometimes in solving a problem, it becomes apparent that some of the data are wrong or inconsistent so that a reasonable solution is impossible or at least uncertain. This should be made clear to the person assigning the problem. This is particularly important in an industrial situation because other people may be using the same data to make important decisions. If possible the correct data should be found or assumed and the solution continued as with too little data.

Algebra before numbers

In problem solving, it is generally quicker to do as much algebra as possible before substituting numbers. Often one will find that some quantities will cancel. Furthermore, by doing all of the numerical operations at one time, there is less chance for error.

EXAMPLE 21.9. There are two ways of finding the percentage of increase or decrease of material costs if magnesium were substituted for aluminum in a part of 2.5 in^3.

Data	Aluminum	Magnesium
Cost/lb	$ 0.86	$1.68
Density	2.70 Mg/m^3	1.74 Mg/m^3

The long method is to calculate the material cost of each part and compare:
The weight of aluminum would be

(2.5 in^3)(0.0254^3 m^3/in^3)(2.7 × 10^6 g/m^3) = 110.6 g or
(110.6 g)/(254 g/lb) = 0.243 lb and this would cost
(0.243 lb)($0.83/1b) = 20.9¢.

The weight of magnesium would be

(2.5 in^3)(0.0254^3 m^3/in^3)(21.74 × 10^6 g/m^3) = 71.3 g or
(71.3 g)/(254 g/lb) = 0.157 lb and this would cost
(0.157 lb)($1.68/lb) = 26.4¢.

Substituting magnesium would increase the cost by a factor of 26.4/20.9 = 1.26 or a 26% increase.

The short method:

$Cost = (Cost/wt)(wt/vol)(vol)$

$Cost_{Mg}/Cost_{Al} = [(cost/wt)/_{Mg}(cost/wt)_{Al}][density_{Mg}/density_{Al}]$

$(vol_{Mg}/vol_{Al}) = (\$0.86/\$1.68)(1.74/2.70) = -1.26$ or a 26% increase.

Note that the conversions of cubic inches to cubic meters and of pounds to grams are unnecessary.

Ratios

Many problems can be solved most easily by setting up ratios. The Arrhenius equation relates the rate of a reaction to temperature. Rate = $A \exp[-Q/(RT)]$ so the time required to reach a certain stage of reaction is given by time = $C \exp[+Q/(RT)]$. Suppose we know two combinations of time and temperature, (t_1, T_1) and (t_2, T_2), that result in the same extent of reaction. We can find the activation energy by setting up the ratio $t_2/t_1 = \{C \exp[+Q/(RT_2)]\}/\{C \exp[+Q/(RT_2)]\} = \{\exp[+Q/(RT_2)]\}/\{\exp[+Q/(RT_1)]\} = \exp[+Q/(RT_2) - Q/(RT_1)] = \exp[(Q/R)(1/T_2 - 1/T_1)] : Q/R = \ln(t_2/t_1)/(1/T_2 - 1/T_1)$

or simply $Q = R\ln(t_2/t_1)/(1/T_2 - 1/T_1)$. Note that it is not necessary to know the value of the constant, C. One can find the temperature, T_3, necessary to obtain the same amount of reaction in a different time, t_3, from $1/T_3 = 1/T_2 + (R/Q)\ln(t_3/t_2)$.

Percentage changes

A percentage change is always defined as $100 \times$ (the difference)/(old value). Specifically, if the old value is x_0 and the new value is x_1, the percentage change in x is $100(x_1 - x_0)/x_0$. This may be simplified to $100(x_1/x_0 - 1)$. With this definition, a 15% increase in price followed by a 15% decrease is less than the original price.

One is always free to define new variables. If, for example, the expression $E/(1 - v^2)$ occurs frequently in a calculation, defining $E' = E/(1 - v^2)$ will reduce the amount of writing.

The use of subscripts may also be used to identify new variables or particular values of a variable. For example, r_o might designate an initial value of a radius and r_1 might designate the radius at some stage 1, or d_i and d_o might designate the inside and outside diameters of a pipe.

Finding slopes of graphs

With two known points, (x_1, y_1) and (x_2, y_2), on a straight line determine that its slope is

$$(y_1 - y_2)/(x_1 - x_2).$$

If there are more than two points on the line, it does not matter which two points are chosen to find the slope. However, experimental data rarely lie exactly on a straight line. There tends to be scatter. A slope that best fits the line can be found by statistical analysis of the points. Without use of such analyses, a good estimate can be found by graphing the data and drawing the line that best represents the data.

If no plot is made, the chance of finding a reasonable approximation is highest if one analyzes two points that are as far apart as possible. Consider the following data, which are plotted in Figure 21.6.

x	y
28	1.0
45	4.15
73	4.9
103	7.9
125	10.8
152	11.7
165	14.0

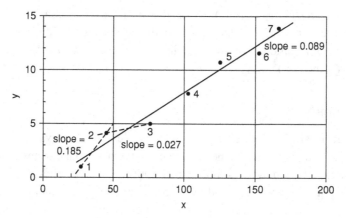

21.6. The best straight line through these points has a slope of about 0.089. Note that the slope between the first two points is $\Delta x/\Delta y = (4.15 - 1.0)/(45 - 28) = 0.185$, and the slope between the second pair of points is $\Delta x/\Delta y = (4.9 - 4.15)/(73 - 45) = 0.0.027$, whereas the slope determined from the extreme points is $\Delta x/\Delta y = (14.0 - 1.0)/(165 - 28) = 0.1095$, which is much closer to the best slope.

Errors in reading a graph are equivalent to scatter of data and have the same effect. If a slope is to be found from a straight line on a graph, reading points well separated from one another will minimize error.

Log-log and semilog plots

There are several reasons for using logarithmic scales. Not infrequently, the values of a quantity being plotted vary by factors of 100, 1000, or more over the range of interest and distinguishing 8 from 10 is as important as distinguishing 800 from 1000. In assessing the growth potential of stocks, it is the fractional (or percent) rate of growth that is important. The price histories of stocks are conventionally plotted on a logarithmic scale. It is the slope on a semilog plot that is important (Figure 21.7).

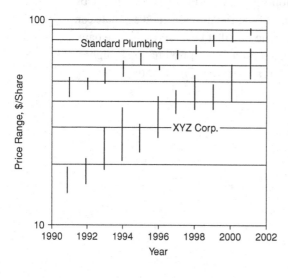

21.7. The change of stock prices over several years. The steeper slope of the XYZ stock indicates a higher percentage of growth, even though the price of both stocks rose by about $50 between 1990 and 2000.

CALCULATIONS

Sometimes there are theoretical reasons for using logarithmic scales. Perhaps it is expected that the data may be described by an equation of the form $y = Ax^n$. In this case $\log(y) = \log(A) + n \log(x)$ so a plot of $\log(y)$ versus $\log(x)$ should be a straight line. There are two different (but equivalent) ways of plotting. One is to calculate the values of $\log(y)$ and $\log(x)$ and plot these. (Figure 21.8A). The other is to plot y versus x on logarithmic scales (Figure 21.8B).

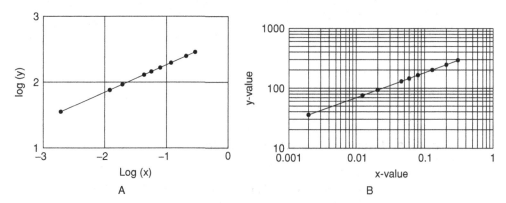

21.8. Two ways of making a logarithmic plot. (A) Either the logarithms of the numbers may be plotted or (B) the numbers plotted on logarithmic scales.

Logarithmic scales are often labeled only at intervals differing by factors of ten with no intermediate grid lines. If x is plotted on a logarithmic scale, the distance between two values x_1 and x_2 depends on the ratio of x_2/x_1. The distance on the paper between 1 and 2 is the same as the distance between 2 and 4 and between 5 and 10. In reading values between 1 and 10 it is well to remember that 2 is at a point about 0.3 times the distance between 1 and 10, so 5 is represented by a point about 0.7 times of the distance between 1 and 10 (Figure 21.9).

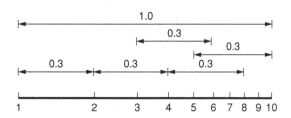

21.9. Reading a logarithmic scale. Note that the paper distance between two points that differ by a factor of 2 is close to 3/10 of the distance between two points differing by a factor of 10.

If the x and y log scales have the same intervals for factors of ten, the slope may be found by measuring Δx and Δy with a ruler and taking the slope, n, by $\Delta y \Delta x$, as illustrated in Figure 21.10A. A more useful way is to take two points $(x_1 y_1)$ and $(x_2 y_2)$. Note that if $y = Ax^n$, $(y_1/y_2) = (x_1/x_2^n)$, $n = \log(y_1/y_2)/\log(x_1/x_2)$. Therefore, one need only to compute the slope from the coordinates of two points. These two approaches are illustrated in Figure 21.10B.

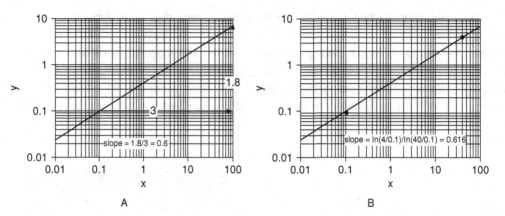

21.10. Two methods of finding the slope of a line on a log-log plot. (A) Using the ratio of the number of decades and (B) using the coordinates of two points.

Graphical differentiation and integration

The differential of a function dy/dx is simply the slope of a graph of y versus x. If the function cannot be expressed analytically, the value of dy/dx at any value of x can be found by drawing a tangent to the curve and finding the slope of the tangent, as illustrated in Figure 21.11.

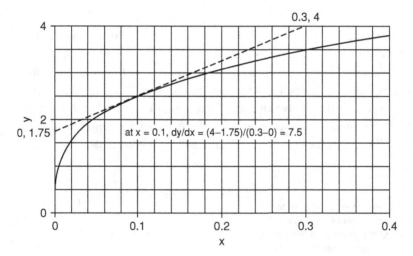

21.11. Differentiation by finding the slope of a graph. The slope, dy/dx, at a point on a curve is the slope of a tangent to the curve at that point. The slope of that line is found by picking two well-separated points on that line. In this case the slope at $x = 0.1$ is found by taking $y = 1.75$ at $x = 0$ and $y = 4$ at $x = 0.3$, so $dy/dx = \Delta y/\Delta x = (4 - 1.75)/(0.3 - 0) = 7.5$.

Integration of a function $y(x)$ between limits of $x = a$ and $x = b$, $\int_a^b y\,dx$, is simply finding the area under a plot of y versus x from a to b. Often the mathematical dependence of y on x is not known or is too complex to integrate analytically. In this case graphical integration may be useful. There are several alternatives.

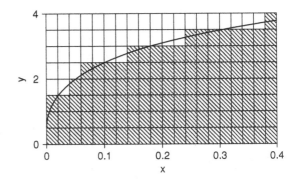

21.12. Graphical integration by counting rectangular elements.

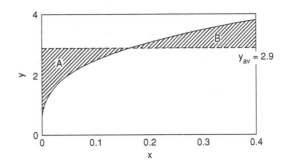

21.13. Graphical integration by eyeballing an average value of y over the interval. A horizontal line representing the average should be drawn so that (A) the area under the curve should equal (B) that over the curve.

One method is to count rectangles of dimensions Δx by Δy under the curve (Figure 21.12). In this case $\int y \mathrm{d}x = n \Delta x \Delta y$, where n is the number of rectangles under the curve. The elements that are completely and those more than 50% under the curve are counted. In the case of Figure 21.12, $n = 57$, $\Delta x = 0.2$ and $\Delta y = 0.5$ so the total area is 5.7. The accuracy of this method increases as the size of the elements decreases.

A rather crude estimate of the integral can be made by eyeballing the average value of y over the range $x = a$ to $x = b$, as shown in Figure 21.12. In this case $\int y \mathrm{d}x = y_{av}(x_b - x_a)$. For Figure 21.13, this method gives $\int y \mathrm{d}x = 2.2 \times 2.5 = 5.5$. This is not a very sophisticated method but it does give a reasonable estimate that can be used as a check for a more sophisticated integration.

The trapezoidal rule provides a rather quick method that is reasonably accurate if the curve is approximated by connecting adjacent points on the curve by straight lines, as shown in Figure 21.14. The interval between each pair of points (x_{n+1}, y_{n+1}) and (x_n, y_n) is a trapezoid of area

$$A_{n,n+1} = (x_{n+1} - x_n)(y_n + y_{n+1})/2.$$

The sum of the areas of all of the trapezoids between $x = n$ and $x = m$ is

$$\int y \mathrm{d}x = \Sigma_{m,n}[(x_{n+1} - x_n)(y_n + y_{n+1})].$$

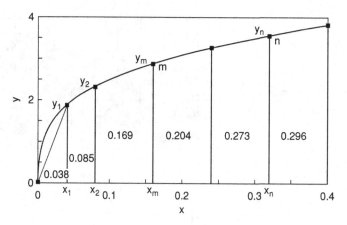

21.14. Numerical integration using the trapezoidal rule. The area under the curve is considered as being composed of a series of trapezoids. The area of each trapezoid is its width, Δx, times its average height, $(y_n + y_{n+1})/2$. The total area is the sum of these areas, $[(x_{n+1} - x_n)(y_n + y_{n+1})]/2$, which in this case equals 5.64. The accuracy of this method is improved by using smaller intervals, Δx, where the curvature is high.

If all of the intervals, $\Delta x = (x_{n+1} - x_n)$, are equal, the summation can be simplified to $\int y\,dx = (x_n - x_m)\Sigma_{m,n}(y_n + y_{n+1})/2$. Or more simply $\int y\,dx = (x_n - x_m)[(y_{\text{first}} + y_{\text{last}})/2 + \Sigma_{\text{2nd, next to last}}(y_n)]$.

This means that one need only find the sum of all points (except the end two), plus $(1/2)$ times the end two, and multiply this by the total x interval. This method works well unless there is appreciable curvature between points. Even then the error is reduced by decreasing the interval Δx.

Iterative and graphical solutions

Many equations cannot be solved analytically. One example is $x^{0.22} \exp(-x) = 0.35$. This equation, however, can be solved numerically by iteration (i.e., trying many values of x and noting how the left-hand side behaves.) The appropriate values of x are those for which the left-hand side $= 0.35$.

x	$x^{0.22} \exp(-x)$	x	$x^{0.22} \exp(-x)$
0	0	0.005	0.3101
1	0.3679	0.010	0.394
2	0.1576	0.008	0.34292
1.1	0.3399	0.009	0.35158
1.05	0.3537	0.0088	0.34992
1.06	0.3509		
1.63	0.35009		

Hence, $x = 1.063$ and 0.0088. We have to decide from the physics of the problem which solution is appropriate. Our original estimate should help us make this decision.

An equivalent method is to let $y = x^{0.22} \exp(-x)$, and plot y versus x. The solution is the value (or values) for which $y = 0.35$. See Figure 21.15.

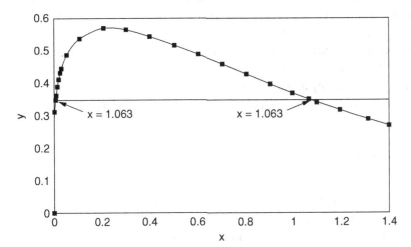

21.15. Graphical solution of $0.35 = x^{0.22} \exp(-x)$. There are two solutions, $x = 1.063$ and 0.0088. One must decide from physical grounds which solution is correct.

Graphical solutions can also be used where some of the information is available only in graphical form.

EXAMPLE 21.10. It is known that the fracture toughness, K_c, of aluminum alloys varies with their yield strengths, as shown in Figure 21.18. In a given situation, fracture will occur if the stress $\sigma = 1.25 K_c$. What yield strength should be specified to allow the highest load without either fracture or yielding? The solution is obtained by plotting $\sigma = 1.25 K_c$ and $\sigma = \sigma_y$ on the same axes and noting the intersection of the two curves (Figure 21.16).

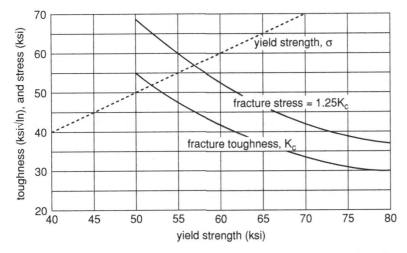

21.16. Graphical solution of the problem of finding the aluminum alloy that will support the maximum load without either yielding or fracture. The intersection at a yield strength of 57 ksi gives the optimum.

Interpolation and extrapolation

Suppose there is a table of data of y at various levels of x and one wishes to know the value of y that is not listed in the table. There are two possibilities. Either the desired value of x lies between two listed values of x, in which case one must interpolate, or the desired value of x lies outside the listed range, so one must extrapolate. Of the two, interpolation is safer.

Interpolation is done by assuming that y varies linearly between x_1 and x_2 so that $y = a + bx$. If the values y_1 and y_2 are listed for x_1 and x_2, $(y_2 - y_1) = b(x_2 - x_1)$ so

$$(y_n - y_1)/(y_2 - y_1) = (x_n - x_1)/(x_2 - x_1) \text{ or}$$
$$y_n = y_1 + (y_2 - y_1)(x_n - x_1)/(x_2 - x_1).$$

This is illustrated in Figure 21.17A. Note that the term $(x_n - x_1)/(x_2 - x_1)$ represents the fractional distance of x_n along the interval x_1 to x_2. This equals the fractional change $(y_n - y_1)/(y_2 - y_1)$ along the interval y_1 to y_2.

The same principle applies to extrapolation, except now $y_n > y_2$. Now

$$y_n = y_2 + (y_2 - y_1)(x_n - x_2)/(x_2 - x_1).$$

This is illustrated in Figure 21.17B.

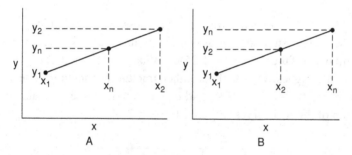

21.17. Interpolating (A) and extrapolating (B).

Analyzing extreme cases (bounding)

It is often useful to calculate an upper bound (a solution that is known to be too high or a maximum possible value). The true solution then is known to be no higher than this. Similarly a lower bound (a solution that is known to be too low or a minimum possible value) sets a lower limit to the true solution. The closer the upper and lower bounds are, the more accurately one can estimate a true solution.

EXAMPLE 21.11. Calculate the number of 1-in.-diameter balls that would fit into a box that is 2 ft × 2 ft × 4 ft.

A simple upper bound can be found by dividing the volume of the box by the volume of a ball. Then $N = (2 \times 2 \times 4 \times 12^3)/[(4/3)\pi \times (.5)^3] = 53 \times 10^3$ is a reasonable upper bound. A lower bound could be found by assuming that each ball occupies a space of 1 in.^{3} In this case $N = (2 \times 2 \times 4 \times 12^3)/1 = 28 \times 10^3$ is a reasonable lower bound. We could make a better upper bound if we happened to know that the maximum possible packing factor for spheres is 74%. With this knowledge the upper bound becomes $0.74 \times 53 \times 10^3 = 39 \times 10^3$. Thus, the true answer lies between 28×10^3 and 39×10^3.

It is often useful to examine extreme cases to see if one is using the right equation.

EXAMPLE 21.12. I have trouble remembering whether the equation for stress relaxation is

$$\sigma = \sigma_o[1 - \exp(-t/\tau)] \tag{21.1}$$
$$\text{or} \quad \sigma = \sigma_o \exp(-t/\tau). \tag{21.2}$$

However, I do know that at time $t = 0, \sigma = \sigma_o$, and at time $t = \infty, \sigma = 0$ so Equation (21.2) is the correct one.

Significant figures

Computers and calculators often give numbers with 7 or 8 figures. Calculated answers are very seldom this accurate because the input data are not known to this accuracy. As a general rule, answers should be reported with as many figures as likely to be accurate. For problems involving only addition, multiplication, and division, this will be the number of figures to which the least accurate input data is known. If subtraction is involved, the accuracy may be much less than the accuracy of the least accurate input. For engineering calculations the accuracy is very often three figures.

When rounding off numbers, a solution should be reported with the greatest number of significant figures so no information is lost but no more. In multiplication, this amounts to reporting the answer to as many significant figures as the least certain input, if the first digit of the answer is higher than the first digit of the uncertain input. Otherwise, one more digit should be reported. The rationale for this is that the relative (percentage) uncertainty in the product is at least as high as the relative uncertainty of the least certain input.

EXAMPLE 21.13. If we calculate $4.032 \times 0.362/8.012 = 0.187207$, the largest relative uncertainty is $\pm 0.005/0.372 = \pm 0.0013$. The absolute uncertainty of the product is $\pm 0.0013x = 0.187207 = \pm 0.0002$, so the product can be rounded off to 0.1872. Information would be lost if we rounded off to 0.187, and reporting the answer as 0.18721 implies more accuracy than is warranted.

On the other hand, for $4.032 \times 0.372/2.731$ there is the same relative uncertainty, $\pm 0.005/0.372 = \pm 0.0013$, but now the absolute uncertainty is $\pm 0.0013 \times 0.5491 = 0.0007$. Therefore, the answer should be rounded off to 0.549.

The temptation to round off numbers before the end of the calculation should be resisted. If all of the calculations are done in a single step this temptation will not present itself because calculators store more figures than the display can show.

EXAMPLE 21.14. Consider the consequence of premature rounding off in the calculation of the ratio D_2/D_1 of the diffusivities of carbon in iron at two temperatures, T_2 and T_1. We know that $D_2/D_1 = \exp[8,900(1/T_1 - 1/T_2)]$. Suppose we want to calculate the percentage of increase of the diffusivity for a 1 °C temperature change at about 500 K. Then $D_2/D_1 = \exp[8900(1/500 - 1/501)]$. If we calculate $1/500 = 1.9960 \times 10^{-3}$ and $1/501 = 0.20000 \times 10^{-3}$ and round these off before we subtract, we would find $1/T_1 - 1/T_2 = 0$ so $D_2/D_1 = 0$ and conclude that there was no change of diffusivity.

If we had not rounded off but had done the entire calculation in one step, we would have found $D_2/D_1 = \exp[8900(1/500 - 1/501)] = 1.035$ or a about a 3.5% increase for each °C.

Rather than writing numbers as 15,000, 0.000,005,5 or 15,300,000,000 it is helpful to the reader to express them in terms of factors of 10^{3n} (e.g., as 15×10^3, 5.5×10^{-6}, or 15.3×10^9). Writing 15.3×10^9 is preferable to 1.53×10^{10}. 10^{3n} reflect the words thousands, millions, 0^{-3} billions, and so on.

When the numbers have units, the factors of factors of 10^{3n} can be expressed by SI prefixes. We can write 15 kJ instead of 15×10^3 J and 3.5 µm instead of 3.5×10^{-6} m. Table 21.1 lists standard SI prefixes.

Logarithms and exponents

There are several simple rules for handling exponents that simplify calculations:

$(x^a)(x^b) = x^{(a+b)}$
$(x^a)/(x^b) = x^{(a-b)}$
$(x^a)^b = x^{ab}$.

Table 21.1. Standard SI prefixes

10^{3n}	Name	Symbol
10^{-15}	femto	f
10^{-12}	pico	p
10^{-9}	nano	n
10^{-6}	micro	µ
10^{-3}	milli	m
10^3	kilo	k
10^6	mega	M
10^9	giga	G

CALCULATIONS

Several simple rules for logarithms are

$$\ln(x^a) = a \ln(x)$$
$$\ln(ab) = \ln(a) + \ln(b)$$
$$\ln(a/b) = \ln(a) - \ln(b).$$

The base of natural logarithms, $e = 2.718$, is defined such that $\ln(e) = 1$.

$$\ln(e^x) = x \ln(e) = x.$$

Note that e^x is often written as $\exp(x)$.

Similarly for logarithms of base 10, $\log(10) = 1$ so $\log(10^x) = x$. Note that $\ln(x) = 2.3 \log(x)$.

The Greek alphabet

Greek letters are often used in science and engineering. Table 21.2 lists some of the common uses.

PROBLEMS

1. Sketch a cube, showing one of the body diagonals between opposite corners. Now calculate the ratio of the length of the body diagonal to the length of an edge.
2. Estimate the mass of the earth.
3. The viscosity of a fluid, η, is defined in terms of a test in which it is sheared. The viscosity is the ratio of the shear stress to the shearing strain rate $\dot{\gamma}$, $\eta = \tau/\dot{\gamma}$. The strain rate, $\dot{\gamma}$, is the rate of shearing between two planes divided by the distance between them. Determine the SI units for viscosity.
4. Plot the y versus x data on the log-log scale given below (Figure 21.18). Determine the exponent n in the equation $y = Ax^n$.

x	y
0.0103	1.256
0.0235	1.497
0.056	1.807
0.113	2.103
0.217	2.398
0.353	2.58

21.18. Figure for Problem 4.

Table 21.2. Greek alphabet

Letters	Symbols	Typical use
alpha	A	
	α	angle, coefficient of thermal expansion
beta	B	
	β	angle
gamma	Γ	mathematical function
	γ	angle, shear strain, surface energy
delta	Δ	difference
	δ, ∂	difference between differential quantities
epsilon	E	
	ε	strain
zeta	Z	
	ζ	
eta	H	
	η	viscosity, efficiency
theta	Θ	*temperature*
	θ	angle, temperature
iota	I	
	ι	
kappa	K	
	κ	
lambda	Λ	
	λ	wave length
mu	M	
	μ	coefficient of friction, shear modulus, 10^{-6}
nu	N	
	ν	frequency, Poisson's ratio
xi	Ξ	
	ξ	
omicron	O	
	o	
pi	Π	
	π	3.14167, ratio circle's circumference to diameter
rho	P	
	ρ	density, radius of curvature, resistivity
sigma	Σ	summation
	σ	stress, conductivity, standard deviation
tau	T	
	τ	shear stress
upsilon	Y	
	υ	
phi	Φ	
	ϕ	angle
chi	X	
	χ	
psi	Ψ	angle
	ψ	
omega	Ω	ohm
	ω	angular frequency

CALCULATIONS

5. From the plot below (Figure 21.19), determine the slope, $d(\ln y)/d(\ln x)$.

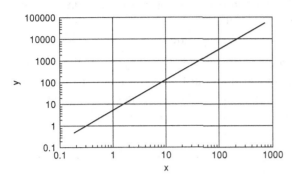

21.19. Figure for Problem 5.

6. The plot below (Figure 21.20) shows the annual production of aluminum beverage cans in the U.S. Find the total number of cans produced between 1970 and 1982.

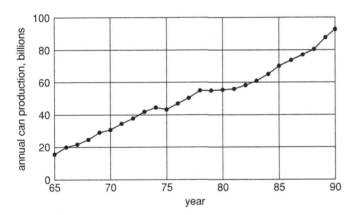

21.20. Figure for Problem 6.

7. The linear coefficient of thermal expansion of aluminum is 23.6×10^{-6}/K. What is the percentage of volume change when aluminum is cooled from 100 °C to 20 °C?

8. From the table below, find the value of x when $\text{erf}(x) = 0.632$.

x	erf(x)	x	erf(x)
0.0	0.0	0.50	0.5202
0.10	0.1125	0.60	0.6039
0.20	0.2227	0.70	0.6778
0.30	0.3286	0.80	0.7420
0.40	0.4284	0.90	0.7970

9. Consider a composite made from plastic resin reinforced by glass fibers. The glass has a density of 2.3 g/cm³ and the resin has a density of 0.95 g/cm³. If the glass fibers occupy 45% of the volume, how many pounds of resin would be required to make 4 in³ of composition?

10. A certain iron-base alloy contains 5% Cr and 10% W by weight. It is desired to make a new alloy with molybdenum substituting for tungsten on an atom for atom basis. That is, one atom of Mo replaces one atom of W. What wt.% Mo should the alloy contain? The following data are available:

Element	Atomic weight (amu)	Density Mg/m³
Fe	55.85	7.87
Cr	52.0	7.19
W	183.9	19.3
Mo	95.9	10.2

Index

alnico, 198
amorphous materials, 153–166
antiphase domain boundaries, 65
asbestos, 176
ASTM grain size, 1
austenite, 58
Avrami kinetics, 108–111

Bloch walls, 191
bonding, 133–142
   covalent, 136
   ionic, 133–134
boundary layer, 94
buckyballs, 180
Burgers vector, 37

carbon fibers, 180
carburization, 80
cementite, 57
characteristic ratio, 155
cholesteric liquid crystals, 168
Clausius–Clapeyron equation, 57
clay, 177
columnar grains, 90
columnar liquid crystals, 169
constitutional supercooling, 95
coordination, 136–139
crystal structures, 140–142
crystal systems, 11
critical radius for nucleation, 86
cube texture, 196
Curie temperature, 66, 171, 191

decarburization, 79
dendrites, 90, 95–98
devitrification, 162
diamond, 179

diffusion, 69–81
   Darken's equation, 77
   mechanisms, 73
   multiphase systems, 78–81
   self, 76
   special paths, 76
   temperature dependence, 75–76
disclinations, 170
dislocations, 36
   energy, 38
   stress fields, 38–39
   partial, 39
distribution coefficient, 92

elastic moduli, 134
enthalpy of mixing, 52
entropy of mixing, 53
error function, 71
estimates, 214–215
Euler relations, 6
eutectic freezing, 98–100
eutectoid transformation, 106–108
exchange energy, 185

ferromagnetism. *See* magnetism
Fibonacci series, 16
Fick's laws, 69–70
foams, 202–204
fractals, 17
free energy, 52
free energy curves, 55–56
free volume, 154
freezing
   cellular growth, 96
   growth, 89
   segregation, 91–93
   single crystal growth, 98

freezing (*cont.*)
  steady state, 95
  volume change, 85
Frenkel defects, 34
fullerenes, 180

gas porosity, 98
gas solubility, 98
Gibbs free energy. *See* free energy
Gibbs, Willard, 49
grain boundaries, 125–127
  area per volume, 3
  wetting, 130
grain size, 1–3
  ASTM, 1
  linear intercept, 1
glass
  bridging oxygens, 158
  chalcogenide, 163–166
  compositions, 158
  delayed fracture, 163
  devitrification, 162
  inorganic, 157–166
  metal, 164–166
  silicate, 157
  thermal expansion, 161
  viscosity, 159
  Vycor, 161
glass transition, 153–166
  polymers, 154
golden ratio, 16
Goss texture, 196
grain boundary, low-angle, 39
graphite, 179
graphical differentiation and integration, 224–226
Greek alphabet, 231
growth, freezing, 89
growth of precipitates, 111–113

hard sphere model, 155
honeycomb structures, 204
hot isostatic pressing, 151

ice, 49, 57, 61
icosahedron, 16
interpolation and extrapolation, 228
interstitials, 33
invariant reactions, 44
ionic radii, 139–140
isothermal transformation diagrams, 108
iterative solutions, 226–227

Johnson and Mehl equation, 108

Kelvin, Lord, 8
Kirkendall effect, 74–75

Le Chatelier's principle, 57
liquid crystals, 168–174
  displays, 174
  orientation parameter, 169
  optical response, 173
  phase changes, 172
  temperature and composition effects, 171
lodestone, 184
log-log and semi-log plots, 222
logarithms and exponents, 230–231
lyotropic liquid crystals, 171

magnetic materials, 184–201
  B–H curve, 190
  hard, 197–198
  oxides, 192
  soft, 194–196
  square loop, 199
magnetic units, 189
magneto-crystalline energy, 188
magnetostatic energy, 187
magnetostriction, 189
martensitic transformations, 114–116
melting points, 134
metal glasses, 164–166
metastability, 57
mica, 176
microstructural relations, 5–6
Miller–Bravais indices, 21–23
molecular length, 154–155

nanotubes, 181
negative thermal expansion, 205
negative Poisson's ratio, 205
nematic liquid crystals, 168, 174
nucleation
  homogeneous, 85–88
  heterogeneous, 88–89
  solid state, 104

order, 64–67
  long range, 64–67
  short range, 67
Ostwald ripening, 113

Pauling, Linus, 142
pearlite, 107
percent changes, 221

INDEX

peritectic freezing, 100
Pfann, William, 93
phase rule, 43–44
plasticizers, 154
point defects, 33–36
porous materials, 202–204
precipitation-free zones, 113
precipitates
   transition, 113
   growth, 111–112
pressure effects, 57

quartz, 178
quasicrystals, 14–17

ratios, 220
radii ratios, critical, 136–138

Scheil equation, 91
Shottky defects, 34
segregation to surfaces, 127
segregation during freezing, 91–93
significant figures, 229–230
shape memory, 208–213
silicates, 176–178
silicon iron, 195
sintering, 144–151
   activated, 150
   early stages, 146
   final stage, 147
   intermediate stage, 147
   liquid-phase, 150
   mechanisms, 144
sketches, 215–217
slopes, 221–222
smectic liquid crystals, 168

solidification. *See* freezing
solubility limits, extrapolation, 60
space lattices, 11
spinodal reactions, 116–118
spherical triangles, 31
stacking faults, 39
steric parameter, 155
stereographic projection, 26–30
superelasticity, 209
surfaces, 121–131
surface energy
   amorphous materials, 125
   direct measurement, 128–129
   magnitudes, 131
   relation to bonding, 121
   relative, 129
   orientation dependence, 122–124

ternary phase diagrams, 44–49
tetrakaidecahedron, 6–8
thermal shock, 160–161
transformations, 104–119

units, 217–219

vacancies, 33
volume fraction phases, 4
Voroni cells, 157

wetting, grain boundaries, 130
Wigner–Seitz cells, 157
whiskers, 40
wulff plot, 123–124

Zachariasen's rules, 157
zeolites, 182
zone refining, 93